Environmental Benchmarking

Becoming Green and Competitive

Francisco Szekely
Thomas Vollman
Annette Ebbinghaus

Business and the environment – practitioner series
Series editor: Ruth Hillary
Centre for Environmental Technology
Imperial College of Science and Technology

STANLEY THORNES

First published in 1996 by:
Stanley Thornes (Publishers) Ltd
Ellenborough House
Wellington Street
CHELTENHAM
Glos. GL50 1YW
United Kingdom

96 97 98 99 00 / 10 9 8 7 6 5 4 3 2 1

A catalogue record for this book is available from The British Library.

ISBN 0 7487 1848 6

Printed and bound in Great Britain by CIP, Basildon, Essex

TITLES AVAILABLE IN THIS AND OTHER SERIES

On this and the following page is a full list of titles in our TQM Practitioner Series, Business and the Environment Practitioner Series and our new Business Performance Improvement Practitioner Series. Some reviews of individual titles are included below.

'...one of the best (reports) yet encountered.'
Automotive Engineer

'...a welcome contribution with lots of useful and clearly written information and guidance.'
Integrated Environmental Management

'...clearly written...useful for anyone in business.'
Waste and Water Treatment

Details of where to get further information can be found overleaf.

Title	ISBN	Price (£)
TQM Practitioner Series		
Writing a Quality Manual for BS EN ISO 9000: 1994	0 7487 2421 4	35.00
TQM – The Hands-On Experience: People Make Quality	0 7487 1845 1	35.00
Forms and Other Documents for Use in BS EN ISO 9000 Quality Systems	0 7487 1846 X	35.00
Essential TQM Tools and Techniques: A Practical Business Handbook	0 7487 1847 8	35.00
Gaining and Benefiting from ISO 9000 Registration: A Step-by-step Guide (Second Edition)	0 7487 1837 0	35.00
A Quality Auditing Manual (Second Edition)	0 7487 1840 0	35.00
Sustaining Quality Advantage in Financial Services	0 7487 1828 1	35.00
Beyond the BS 5750/ISO 9000 Certificate: The Bureaucracy Buster's Guide to Quality Assurance	0 7487 2106 1	35.00
Writing an ISO 9000 Quality Manual: Practical Guidelines	0 946655 89 8	35.00
Risk Management in a TQM Environment	0 7487 1905 9	35.00
FMEAs: A Manager's Handbook	0 7487 1861 3	35.00
TQM: Its Impact on Bottom Line Results	0 946655 96 0	35.00
The Role of Junior and Middle Level Management in TQM	0 946655 74 X	35.00
Quality Function Deployment	0 946655 72 3	35.00
The Role of the Modern Quality Manager	0 7487 1867 2	35.00
Measuring Service Quality	0 946655 75 8	35.00
TQM in Sales and Marketing	0 946655 76 6	35.00
TQM: Does It Always Work?	0 7487 1936 9	35.00
Communication in Successful TQM	0 7487 1872 9	35.00
Kaizen	0 7487 1901 6	35.00
TQM Training	0 7487 1871 0	35.00
Teams and TQM	0 7487 1870 2	35.00
QA and ISO 9000/BS 5750 in the Hospitality and Catering Industries	0 7487 1869 9	35.00
BS 7750: What the New Environmental Management Standard Means for your Business	0 946655 60 X	35.00
TQM in Service Industries: A Practitioner's Manual	0 7487 1914 8	35.00
Managing and Improving Service Quality and Delivery	0 7487 1907 5	35.00
Competitive Benchmarking: An Executive Guide	0 7487 1926 1	35.00
TQM-Based Performance Measurement: Practical Guidelines	0 7487 2594 6	35.00
Quality Costs: Their Impact on Company Strategy and Profitability	0 7487 1875 3	35.00
Customer Care: Strategy for the 90s	0 7487 1868 0	35.00
Ensuring Your Business Achieves and Profits from BS 5750 Registration: A Practical Guide	0 946655 44 8	35.00
Leadership of Customer-Driven TQM	0 7487 1865 6	35.00
TQM in the Electronics Industry: A User's and Supplier's Guide	0 7487 1874 5	35.00
Software Quality Assurance: What It Buys You and What It Costs You	0 7487 1902 4	35.00
Implementing TQM in Small and Medium-Sized Companies	0 946655 59 6	35.00
Selling Professional Services	0 7487 1873 7	35.00

Continued overleaf

Business and the Environment – Practitioner Series		
Environmental Benchmarking: Becoming Green and Competitive	0 7487 1848 6	35.00
Managing Environmental Crises	0 7487 2439 7	35.00
The Role of the Environmental Manager	0 7487 1842 7	35.00
Environmental Compliance Auditing	0 7487 2187 8	35.00
Measuring Environmental Business Performance: A Comprehensive Guide	0 7487 2063 4	35.00
Environmental Improvement Through the Management of Waste	0 7487 2129 0	35.00
The 1994 Waste Management Licensing Regulations and the New Definition of Waste	0 7487 2133 9	35.00
International Environmental Policy and Management	0 7487 2132 0	35.00
Integrated Pollution Control	0 7487 2061 8	35.00
Life Cycle Assessment	0 7487 2131 2	35.00
The Implications of Water Regulation for Industry	0 7487 2062 6	35.00
A Trouble Shooter's Guide to Environmental Management Systems: BS 7750: 1994 and Other Options	0 7487 1841 9	35.00
The Eco-Management and Audit Scheme: A Practical Guide	0 7487 1900 8	35.00
Business Success from Seizing the Environmental Initiative	0 946655 93 6	35.00
Improving Environmental Performance: A Proven and Effective Approach	0 7487 1864 8	35.00
Managing Environmental Risks and Liabilities	0 7487 1906 7	35.00
Contaminated Land: Assessment and Redevelopment	0 7487 1937 7	35.00
The Environment and the Planning System: Business Implications	0 7487 1866 4	35.00
Business and Environmental Accountability: An Overview and Guide to the Literature	0 7487 1863 X	35.00
Environmental Auditing: A Guide to Best Practice in the UK and Europe	0 7487 1862 1	35.00
Directors in the Dock	0 7487 1927 X	35.00

Business Performance Improvement – Practitioner Series		
Getting the Right Product	0 7487 2453 2	35.00
Writing Effective Performance-Related Procedures	0 7487 2503 2	35.00
Corporate Communication and Business Performance	0 7487 2572 5	35.00
Quality In Business Performance: How to Change Attitudes	0 7487 2438 9	35.00
Training Industrial Staff: A Practical Handbook	0 7487 2369 2	35.00
Communicating for Improved Business Performance	0 7487 2266 1	35.00
Getting Closer To Your Suppliers	0 7487 1852 4	35.00
Customer Satisfaction Measurement: A Practical User's Guide	0 7487 2138 X	35.00
Competitive Customer Service	0 7487 2135 5	35.00
Getting Good Value from Consultants	0 7487 1851 6	35.00
Commercial Awareness: Increasing Your Edge	0 7487 1838 9	35.00
Business Control, Accountability and Corporate Governance	0 7487 2134 7	35.00
An Implementation Guide on How to Re-engineer Your Business	0 7487 2130 4	35.00
Practical Implementation of New Management Strategies: Managing Your Processes	0 7487 1839 7	35.00
Ensuring Your Business Survives: A Self Analysis and Self Help Manual for Managers	0 7487 1844 3	35.00
Self Assessment: A Tool for Integrated Management	0 7487 2060 X	35.00
Better Planning for Performance Improvement	0 7487 1853 2	35.00
Changing the Business: A Practical Guide to Performance Improvement	0 7487 1822 2	35.00
Business Process Re-engineering: A Practical Handbook for Executives	0 7487 1821 4	35.00

If you would like further information, please contact us at the address below. Many companies buy bulk copies of our reports; if you are interested in doing this, we would be pleased to discuss discount levels.

Customer Services Department
Stanley Thornes (Publishers) Limited
Ellenborough House
Wellington Street
CHELTENHAM
Glos. GL50 1YW

Telephone: (01242) 228586
Fax: (01242) 221914

ABOUT THE AUTHORS

Francisco Szekely

Dr. Francisco Szekely works with McKinsey and Co. as a member of the firm's environmental management practice. He graduated in Engineering from the National Autonomous University of Mexico, and gained an MS from Tufts University, USA. He holds an MS in Technology and Human Affairs, and a PhD in Environmental Science and Economics, from Washington University, USA. Formerly, he was ALCAN Professor of Environmental Management at IMD, and was the founder of the MIBE Project. He also served as Assistant Director of the Energy and Environmental Policy Center at Harvard University and has lectured on the environment and economic development at Harvard University and on environmental planning at MIT. He currently advises multinational companies on environmental management matters and is Senior Advisor to the UN Environmental Program in Geneva and the UN Development Program in New York.

Thomas E. Vollmann

Thomas Vollmann is Professor of Manufacturing Management at the International Institute for Management Development (IMD) in Lausanne, Switzerland, where he also serves as Associate Director of the research project Manufacturing 2000. He was educated at the University of California, Los Angeles, returning after the undergraduate degree to earn both an MBA and a PhD. Prior faculty positions include Boston University, INSEAD, Indiana University, University of Rhode Island, and Dartmouth College. Professor Vollmann's areas of special interest are manufacturing planning and control, performance measurement, supply chain management and enterprise transformation. He is the author or co-author of eleven books, numerous journal articles, and is currently leading an IMD research project, Manufacturing Transformation, which is described in the book, *The Transformation Imperative: Achieving Market Dominance Through Radical Change*. He is actively involved in executive education and consulting.

Annette Ebbinghaus

Annette Ebbinghaus is a Research Associate with the Alliance for Global Sustainability. She holds a BASc in Civil Engineering from the University of Windsor, Canada, a MASc in Civil Engineering from the University of Ottawa, Canada, and an MBA from the Western Business School, University of Western Ontario, Canada. Prior to working at IMD she consulted as a structural engineer to both the private and government sectors. As a Research Associate on IMD's MIBE Project she was responsible for managing several environmental benchmarking studies, researching industry best practice and investigating how companies build organisational and management systems to support 'green champions'. She developed and is currently publishing a framework for environmental integration based on best environmental practice coupled with current charters and principles for governmental and environmental organisations in North America and Europe. She is also on the editorial board of *Environmental Excellence Magazine*.

ACKNOWLEDGEMENTS

This book is largely based on the environmental benchmarking activity developed in connection with IMD's Managing the Industrial and Business Environment (MIBE) project. We are indebted to the companies which were willing participants in this 'active learning' partnership. A special acknowledgement goes to Sulzer Technology Corporation for sharing their experience in environmental benchmarking both with us - and with you, the readers of this book. Additionally, many participants in IMD's executive education programmes have helped us refine our ideas as we have applied the environmental business transformation diagnostic; we thank them for their valuable feedback and for helping us see its application in additional industry sectors.

The 'Environmental' Business Transformation Diagnostic evolved from a transformation diagnostic process originally developed in IMD's research project, Manufacturing 2000 (M2000). We wish to acknowledge both the companies in M2000 and the project team which developed and perfected this diagnostic process.

There are several additional people we wish to mention. We would first like to acknowledge the work of Faith Towle who has been serving as our editor on this project. Her input has been invaluable in questioning our ideas as well as in bringing three differing approaches and writing styles into harmony. We also thank Dr. Peter Lorange, President of IMD, for his continued support for both the MIBE and M2000 Projects and for this writing project.

In closing we would like to acknowledge the special appreciation we owe our spouses. Sally, Tani and Dan have given their continued support and encouragement as we toiled away nights and weekends to bring this piece to fruition. It is to them that this book is dedicated.

TABLE OF CONTENTS

Chapter One

THE ENVIRONMENTAL IMPERATIVE

Modern society demands a great deal from its business leaders. Often these demands seem irreconcilable, including the need to maintain or achieve excellent economic results while at the same time taking a major role in protecting the environment. Increasingly, however, achieving these apparently conflicting demands is proving to be not only possible - but is being used to obtain a critical source of competitive advantage.

Indeed, the answer to this conundrum - at least in the abstract - is that companies can achieve excellent economic performance while improving environmental effectiveness, but it is not easy.

Objective Our primary objective is to provide a path for managers to incorporate environmental dimensions into their daily business practices in ways that allow them to conserve the environment *and* to increase their competitive ability over the long term. We also intend to provide information, methodologies and tools that can measure the environmental performance of manufacturing processes, service operations and commercial activities.

* **The environmental imperative** The fundamental message of the book is that it is possible to carry out environmentally-sound business development which is cost effective and competitive. Moreover, those firms which do so will dominate their industries, while those which see environmental issues as being secondary or only a form of good citizenship will lose out.

The approach of the book is pragmatic. Our knowledge is derived from the experiences of business managers who have learned - mostly by trial and error - to deal with the environment in an effective and efficient manner. This experience base provides lessons that others can follow: it is through the sharing of best practices that other managers can foster their own learning processes for this critically important subject.

Benchmarking The methodology we propose is 'benchmarking', a management tool which we shall explicitly apply to environmental improvements that will increase competitiveness. This methodology has proven to be very useful, enabling companies to compare their environmental practices with those of others that have achieved 'excellent environmental performance (EEP)'. That is, companies are becoming or remaining business leaders in their industrial sector, when one key part of that leadership is based on leveraging their approach to solving environmental problems.

The present world economic situation demands that companies develop the best strategies they can to differentiate themselves and become more competitive. This book shows that, in order to stay competitive over the long haul, a company must learn to manage its environmental performance in an appropriate manner. A central tenant of this book is that business managers benchmark their environmental performance to improve their business practices and become more competitive.

Benchmarking environmental performance This activity encompasses the comparison of various business activities and practices such as the design of products and services, the development and adoption of cleaner technologies, the elaboration and implementation of effective communication skills, environmental management strategies, and other business activities that implicitly or explicitly affect the environment. But, benchmarking environmental performance is more than a plea to find and copy best practices. This book develops a concrete diagnostic process through which firms have been able to determine a workable environmental strategy - and thereafter follow through with the particular skills and infrastructure required to make this strategy an operational reality.

Benchmarking in an environmental context is not a static activity - it is a continuous process of improvement. The process is *relative* because it aims to identify the best available practices at a given time, according to current technology and economics, even if this practice does not resolve all environmental concerns. At the same time, the process is *dynamic* because environmental targets are continuously changing; what might be identified today as best practice, or specific environmental target, can be different tomorrow, according to the particular experience gained, new available technology and common practices. Thus, we see the adoption of common standards and practices for 'end-of-pipe' issues now augmented with new environmental objectives and measures. And, even more important, it is the environmental practices of leading-edge firms that *set* the environmental standards. That is, Excellent Environmental Performance (EEP) becomes an internal part of *achieving competitive* advantage - and *keeping it.*

The concepts in this book are corroborated by first-hand experiences obtained from interactions with managers from the companies participating in IMD programs and, more particularly, with managers affiliated with companies that have been one of our research partners in the IMD project, Managing the Industrial and Business Environment (MIBE). The companies which have been active in the MIBE project are:

* Alcan Aluminium Ltd.
* Baxter Healthcare Ltd.
* Cementa Euroc AB
* General Electric Plastics bv
* Northumbrian Water Group PLC
* Saga Petroleum AS
* Société Générale de Surveillance SA
* Sulzer Technology Corporation
* Tupperware (Europe, Africa and Middle East)
* Volkswagen AG

Road map The remainder of this book focuses on providing the answers to achieving profitable environmental practice. Chapter 2 describes the environmental movement and various responses from the corporate world. This chapter sets the stage for why environmental benchmarking can be used to increase competitiveness.

Chapter 3 provides the definitive link between benchmarking and environmental improvement; it defines a 13-step benchmarking method, which is divided into four phases: discovery, assessment, action, and continuous improvement. Chapter 3 also introduces a company example that we follow through several chapters.

Chapter 4 deals with *discovery*. The strategic environmental issues are described in detail, with explicit linkages established between the internal/external environmental discontinuities facing the firm, the changes in expectations of customers/other stakeholders, and the strategic responses of the business. The chapter then focuses on the idea of achieving environmental competencies that are explicitly focused to achieve a chosen environmental strategy. We then develop the 'competency map' idea, which determines the detailed infrastructure (systems and processes) needed to generate a competency. In this chapter, we also see how these issues are played out in our example company.

Chapter 5 focuses on *assessment*, where the specific practices and processes for environmental benchmarking are established: the benchmarking team is defined, the sites for benchmarking are selected, the necessary homework is prepared in detail, the way to approach the actual benchmarking visits is described, and common mistakes are identified. This chapter concludes with a description of the example company's experience.

Chapter 6 deals with *action*. We explain the process of actually carrying out the benchmarking study, including how to make the visits, procedures to take for gaining an optimal understanding, and the best way to report results so as to improve existing processes. Following the practice in the preceding chapters, we conclude by relating the experience of our example company in this phase of the environmental benchmarking process.

Chapter 7, *continuous improvement*, explains why the search for environmental excellence - excellence which also produces competitive advantage - is never ending. It is always necessary to ask 'what next'. This chapter also notes that, in the long run, it is the firms which really work at making continuous improvements that achieve a competitive advantage.

Chapter 8 explores the ongoing development of environmental standards and some additional environmental management tools that can be adopted within the benchmarking framework. In particular, the practice of life-cycle assessment is explained - in terms of current best practice, and benchmarking networks are identified.

Chapter 9 discusses how the different social 'actors' are shaping the present environmental agenda. The critical issue is for companies to understand (and anticipate) how the environmental agenda is being shaped, and to know how to respond to the resultant challenges in the most positive fashion.

Finally, Chapter 10 presents messages for managers, learning points and recommendations. Here, the intent is to draw the major conclusions from each of the chapters in a concise way - one that focuses the attention on how to go about achieving the environmental imperative: achievement of best environmental practices is sound business practice and leads to the development of a distinctive competitive advantage.

Chapter Two

THE WORLD RESPONDS TO THE ENVIRONMENTAL MOVEMENT

Although environmental issues have always existed, the present movement started less than 30 years ago after a series of independent incidents highlighted the need to pay increasing attention to the environment. The modern day environmental movement is closely associated with the works of four groups: Society, Governments, International Organisations, and Business. Environmentalists, who are members of all four groups (often more than one) do not form a separate group per se; rather, they strive to push the environmental debate forward by expressing their dissatisfaction with the environmental agendas of each of the four groups. As well, they participate in and give financial support to a number of environmental organisations - such as Greenpeace[1] - which have, in turn, become businesses in their own right. In this chapter, we briefly follow the history of the environmental movement and the responses made by the business world as we answer the following questions:

* What have been the landmark events of the environmental movement?

* How has business responded to the environmental challenge?

* What are the strengths and weaknesses of this response, and how are firms developing a competitive advantage through environmental actions?

The Landmark Events of the Environmental Movement

The environmental movement has its own history, which is largely based on four main events. Indeed, numerous incidents over the years can be seen as logical extensions of these particular landmark events. But, in this context, descriptions of these main events are presented merely as a prelude to the responses made by business.

The landmark event for society In 1967, the American author Rachel Carson wrote the book, *Silent Spring*, which sadly relates why the robin, formerly the herald of spring in North America, had gradually almost disappeared. After conducting some research, Ms Carson learned that excessive use of agricultural pesticides in the Golden Valley of California, the robin's winter habitat, had affected its egg growth rate and thus its population. The book touched the emotions of people everywhere, and her message significantly raised social awareness regarding the need to control potentially hazardous substances and to take more serious steps to protect nature.

The landmark event for generating business awareness At the end of the sixties, a group of well-known business leaders headed by the Italian industrialist, Aurelio Peccei - then

---

[1] Greenpeace was formed in 1971 when members of the 'Don't Make A Wave Committee' in Vancouver, Canada renamed their organisation to better match their purpose: to create a green and peaceful world. Today, Greenpeace is a network of national and regional offices, coordinated from Amsterdam. More than 1,330 people currently staff 43 offices in 30 countries.

President of Olivetti International, became concerned about the potential scarcity of resources worldwide. Their fears were based on the paradigm developed by Malthus in the 19th century - that as population grows exponentially and resources grow arithmetically, eventually the number of people on the planet would exceed the available resources. Malthus also predicted that the period of real scarcity would be preceded by a harsh struggle for access to and control of the remaining resources, which would ultimately lead to a political upheaval, and much death and suffering.

This group, which eventually called themselves 'the Club of Rome', hired Professor Jay Forrester of the Massachusetts Institute of Technology (MIT) to create a model, using his popular 'systems dynamics' technique, for several world scenarios based on the growth of one or more of five basic variables: population, technology, food production, energy demand and pollution. The results of Forrester's modelling, entitled *The Limits to Growth,*[2] predicted that the world would collapse at some future date if the growth of these variables, particularly population, were not stabilised or halted. Many people thus assumed that most environmental problems could be attributed to population growth rates; however, really rapid population growth was only taking place in third-world countries.

The landmark event for international organisations In the summer of 1968, when some Danish fishermen discovered that their catch had been severely affected by industrial contaminants flowing into the water from Sweden, they asked the Danish government to request economic compensation from Sweden. However, at that time no international legal framework existed for resolving problems of this nature.

Motivated by the need to clearly define rules and procedures to handle this type of complex dispute, Sweden initiated an international gathering to discuss environmental problems of general concern. More specifically, the meeting's purpose was to discuss the so-called 'problems of the commons' - environmental problems with global consequences. Responding to this call, the United Nations (UN) organised the United Nations Conference on the Human Environment (UNCHE), which took place in Stockholm in June 1972. The Stockholm conference was a milestone in the environmental movement: for the first time, the role and responsibility of *countries* was acknowledged, which in turn led to environmental legislation as well as to a much more profound understanding of worldwide implications of all environmental issues.

The Stockholm conference also provided another important conceptual breakthrough - a clear link was made between economic development and environmental conservation. Indira Gandhi, representing India (population 800 million) observed: 'Environmental conservation's main enemy is poverty.' In other words, in many societies, people destroy the environment out of need. When people are concerned about how to survive, they are not going to stop and think about conservation. Indeed, the point was made at the Stockholm conference that although many people harm the environment because they are poor and think only of satisfying basic needs, there are many others who contribute equally to environmental deterioration by over consuming resources and misusing their wealth.

In the industrialised countries and the rich segment of developing countries, people consume resources and products that generate waste detrimental to the quality of life, destroying the

---

[2] Meadows, Denis and Donna, *The Limits to Growth*, Cambridge, MA: MIT Press, 1971.

environment for themselves and their less affluent neighbours. For example, the average German citizen in 1972 consumed ten times more energy and mineral resources than his counterpart in Malaysia or Argentina. Germans also generated 13 times more waste than someone from one of the two developing countries. Thus, the Stockholm conference revealed that people destroy the environment for two starkly contrasting reasons: *need and greed.*

As well, the Stockholm conference generated a wealth of information which had never before been available to the scientific community and to decision-makers. In particular, it provided detailed scientific information about the state of the environment, the lack of institutional structures and mechanisms to preserve the environment, and the vacuum of economic and political instruments to create incentives for people to preserve, protect and restore the environment. One immediate result of the Stockholm conference was that national and international environmental policies were rapidly developed, and regulatory schemes and activities to preserve the environment at both local and global levels were designed and adopted.

The landmark event for governmental entities After the 1972 UNCHE conference in Stockholm, individual countries and their governmental entities began the task of developing environmental legislation and policies. The result was the birth of numerous regulations and mechanisms for monitoring their implementation. In this case, no *single* governmental event occurred; rather, many governmental entities agreed to accept responsibility for improving the environment. The ensuing environmental debate then took an important turn in 1986 when, at the urging of the governments of Japan and Norway, the United Nations developed a body known as the World Commission on Environment and Development (WCED). Its purpose was to draw attention to the way that countries were dealing with their environmental responsibilities at home and abroad. The WCED, chaired by Gro Harlem Brundtland, Prime Minister of Norway, produced a 'global agenda for change' in 1988. This report contributed significantly to the environmental cause by launching the concept of 'sustainable development' at the international level. Initially, countries were not united in their approach nor convinced that the more developed countries needed to take more responsibility for ensuring that business conducted in developing countries was at equally sustainable levels as at home. The governmental entity 'event' - or 'non-event' according to some - is critical as it raises an essential issue: individual governments need to act. But, in doing so, their actions will generate two related problems. First, there is a tendency to assign blame or necessary action elsewhere; second, a common focus, rules and approaches need to be developed.

Unified action The four defining events converged in June 1992. Twenty years after the Stockholm Conference, the United Nations organised another conference in Rio de Janeiro in order to assess the progress that had been achieved in the intervening years and to evaluate the status of sustainable development around the world. This conference, known as the 'Earth Summit,' was the largest gathering of the international community since the creation of the League of Nations. Around 35,000 participants from governments, industry, pressure groups and the media went to Rio de Janeiro to express their grievances, hopes and potential solutions for international cooperation.

Unfortunately, the Earth Summit failed to accomplish as much as expected beforehand. There were two major disappointments:

* Although the United Nations had spent two and half years preparing a detailed program - called 'Agenda XXI' - to lessen and correct the main global environmental problems, the government delegates to the Summit were not willing to pay the costs of its implementation. They claimed that a global solution could only be achieved through international cooperation.

* The government delegates also neglected the issue of deforestation. They refused to sign an international convention to protect forest resources, comparable to the international conventions on global warming and biodiversity; instead, they prepared a weak set of principles 'inviting' government authorities to look after the conservation of their forests. These principles were not legally binding nor did they include any particular targets.

Nevertheless, two important international conventions *were* signed at the conference: The Global Warming Convention, designed to reduce carbon dioxide emissions from man made sources; and the Biodiversity Convention, intended to protect flora and fauna around the world, particularly endangered species that need special and immediate protection.

Back to business On the whole, individual business entities have been reluctant to respond to the four landmark events regarding the environment. As we shall describe, this was largely due to the environmental agenda being first defined in command and control terms - where business actions were mandated, and costs were to be incurred without attributed benefits. But this has changed. Since the beginning of the 1990s, many enterprises - and entire industries - have started to acknowledge that environmental responses by business do not need to be seen only in adversarial terms. It is increasingly possible to observe the business benefits derived from improving environmental performance: becoming more cost effective, less risky, more customer focused (in terms of environmental attributes), and more competitive - to the point of gaining dominance in their industry sectors. As well, many governments have begun to issue international, regional and national environmental business standards (see Chapter 8 for more details), which encourage industry to be a key player in solving environmental problems.

From this point on, the book concentrates on the business response to environmental challenges and how it can achieve the environmental imperative: sound environmental business development and competitive advantage. We shall focus particularly on the environmental benchmarking process as it is a valuable tool that business can use to improve both environmental and business performance.

The Business World Responds to the Environmental Challenge

The aftermath of the 1992 conference in Stockholm led to countries developing national environmental legislation and policies. In addition, detailed environmental standards were adopted to preserve different components of the ecosystem (air, water, soil, renewable natural resources) and to protect human health. This first attempt at environmental management, now referred to as the 'command and control' era, met a universally *reactive response* from the business community. Subsequently, three other approaches to improving the environment were tried: first, market-based incentives to obtain a *productive response*; then, industry self-regulation to encourage a *proactive response*; and finally, sustainable development to promote a *visionary response* (see Figure 2.1).

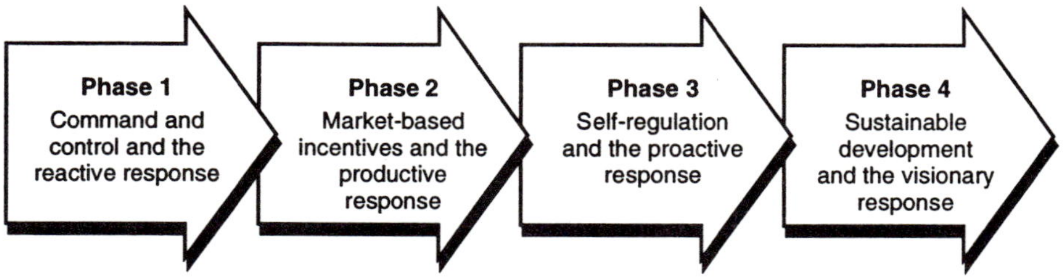

Figure 2.1 Four phases of business response to the environmental challenge

Command and control - and the reactive response The basic assumption behind the command and control approach is that governments must force business to respond to environmental issues by passing laws and ensuring their compliance with those laws. However, the command and control approach is based on the premise that governments are able to determine what represents good and bad environmental policy and what regulations are needed to enforce their compliance. The success of using this approach to provide an improved environment depends on four key assumptions:

* Sufficient scientific knowledge exists to determine what is and is not acceptable for the environment, i.e., the amount and type of pollutants entering the environment will not alter it in an irreparable manner.

* A good monitoring program is established to identify any changes inflicted on the environment over time.

* An effective and efficient enforcement program is in place to penalise those who transgress the limits established by the law.

* The rules generated by the command and control approach are both 'practical' and able to adjust for 'innovation'. That is, they are technically possible, they *can* be achieved, and they can be fulfilled at a cost level that does not preclude competitive viability.

In putting the command and control method into practice, people in government positions who believed in improving the environment were eagerly assisted by environmental pressure groups, in defining standards and helping to enforce compliance. As a result, many environmental policies, laws, standards and institutions were created.

The reactive response shown by business to this approach was not at all surprising. Most companies felt under attack, and they were also afraid as they did not know where these measures were headed. Inevitably, many jokes about ridiculous regulations began to circulate. Fundamentally, the situation was adversarial, confrontational, and each incident typically resulted in a negotiation process. The industry attitude reflected two important issues: first, companies were not convinced that the environmental demands were based on hard scientific evidence, and that such demands were either unjust or unnecessary both for them and for society as a whole. They felt that the costs of complying with these demands was a waste of money that would make their products more expensive and eventually less competitive. Secondly, many companies considered the environmental movement to be only a 'fad' that would fade away with time.

Among those expressing a reactive response were the producers and consumers of chlorofluorocarbons (CFCs). These companies argued that until there was enough scientific evidence to prove a direct relationship between production/consumption of CFCs and depletion of the ozone layer, they would not stop their activities. This attitude lasted approximately 13 years. In 1985, a detailed analysis of NASA data confirmed that there was a definitive correlation between CFCs and the diminishing ozone layer. The resultant work of the Vienna Convention for the Protection of the Ozone Layer and the subsequent work of the Montreal Protocol on Substances that Deplete the Ozone Layer were adopted by the international community. This research acted as a mandate to governments to force companies to stop producing and consuming CFCs. The regulation went into effect on January 1, 1995 for industrialised countries and Mexico (which volunteered to join the industrialised countries), and ten years later for most developing countries.

Another more recent example of the use of command and control - followed by a reactive response - occurred in the packaging industry in Europe and North America. Many pressure groups concerned about the large amounts of waste generated by consumers, as well as the economic and environmental costs that such waste incurred on society, demanded legislation to ban the production of one-way packaging. They called for reusable packaging as well as the reuse and recycling of packaging materials. In this case, companies were right in asserting that there was not enough scientific evidence regarding the environmental impact of different packaging materials.

The problem is a complex one; it is not just a simple matter of comparing one type of packaging with another - for example, using plastic crates versus one-way cardboard transport packaging. Plastic crates need to be cleaned, they are more expensive to store, they are more easily damaged unless additional packaging is used to keep the items firmly in place inside the plastic containers, as well as a host of other issues - such as corrugated boxes offer an excellent use for recycled paper. For every life-cycle assessment (LCA) that shows the benefits of packaging materials such as plastic over paper, glass or aluminium, there are at least two other LCA studies that conclude precisely the opposite. Neither legislators, academics nor pressure groups have been able to obtain the 'hard scientific' evidence necessary to settle this dispute. The packaging debate is a subject that most people believe they can understand. Everyone sees packaging - and the waste; Europe alone produces 500 million tons of packaging related waste per year. Much of this waste may indeed be unnecessary, and many claim that the packaging companies should be responsible for dealing with it. They need to work on avoiding excessive packaging in the first place *and* taking it back for further reconditioning and use. Many companies in this industry are still taking a reactive approach, but others are not. New product development and recycling of packaging materials are both increasing. The debate is especially relevant to companies like Tetra Pak, where packaging is its one and only product and thus its survival as a business is at stake.

The command and control approach to improving the environment has had limited success, when judged in terms of costs and benefits. For example, between 1970 and 1990, the United States spent $1,000 billion developing and enforcing programs and activities to comply with air pollution standards.[3] However, the results have been rather poor. Although some air pollution parameters such as sulphur dioxide and nitrogen oxides have been substantially lowered, other important pollutants - carbon dioxide and hydrocarbons - have increased significantly during the same period.

There is a fundamental problem with command and control as a mechanism to improve the environment. This approach creates an adversarial relationship between government units and pressure groups, which see businesses as polluters to be reined in, and companies that see environmental policymakers as 'the enemy' with no understanding of the realities of running a competitive business. Thus, the command and control approach almost always leads to a reactive response - and, typically, a vicious circle.

Market-based incentives - and the 'productive' response In recent years, there has been a move away from the command and control method to other approaches - such as using the market - to protect and conserve the environment. Market-driven incentives include:

* Introducing economic schemes, particularly market mechanisms, which will optimise the allocation of society's resources. With this approach, pollution rights and resources are traded like any other good in the market until an optimal price for environmental quality is found between users and sellers;

* Developing new products and services that are 'environmentally-friendly' and allowing consumers to express their environmental preferences through the marketplace;

* Using governments, consumer groups and other citizen associations to develop environmental management standards designed to promote competitiveness.

The critical difference between command and control and market-based incentives is that the former designates 'those who know what is good for the environment and those who don't', which automatically leads to confrontation and a reactive response. In the market-based approach, this attitude is largely replaced by an explicit understanding that more global issues need to be addressed - and within an economic context. If company A can reduce its air pollutants below what is mandated, it should be able to sell the extra reduction to company B. The overall air quality would improve, and the costs of non-compliance would become established by the marketplace.

Some companies have discovered that they not only have an obligation as members of society to improve the environment, they can reap economic gains from these activities and even dominate their industry sectors by defining the government standards for environmental performance. That is, by accommodating society's environmental demands, they can contribute to protecting the environment, generate business by producing highly desirable 'environmentally-friendly' products and services, and be among the first to address the environmental issues pertaining to their industry sector.

---

[3] Commoner, Barry, "Can Capitalists Be Environmentalists?", *Business and Society Review*, November, 1993.

The German company Henkel AG is a good example. Management clearly understood that one of their main products, detergents, was responsible for increasing the total phosphate concentration in German water ecosystems, which eventually led to the eutrophication[4] of German water bodies. Henkel developed a phosphate substitute, Zeolites, which it tested for acceptance and began to use in its detergents. In doing so, Henkel contributed to environmental conservation - and gained a larger market share until other competitors emulated and adopted this technology. Henkel's actions also influenced the German authorities to establish a policy that all detergents produced and consumed in Germany thereafter be phosphate free. Thus, in this case, it was the *company* that established the standard - based on *its* technological innovation - rather than the government.

The company, Body Shop, offers another example of a 'productive' response. Anita Roddick, an Englishwoman who was concerned that cosmetic producers were using unnatural substances in cosmetics and testing their products on animals, decided to develop a new set of environmentally-friendly and socially-conscious products. All Body Shop products, which contain only ingredients derived from organic sources, are tested exclusively on bacteria and human volunteers, never on animals.[5] Her ideas appealed to many consumers who decided to support her as a means of contributing to the environment, to animal rights, and to sustaining the livelihood of the local people in the developing countries where some of the ingredients originated. Ms Roddick started her company at home in 1973 with only $25,000 in capital. Her company has grown significantly ever since, both in the UK and abroad. In 1994, Body Shop had sales of $493 million, in 43 countries.

These two examples illustrate the use of the 'productive' approach to the environment. Fundamentally, these companies used an economic motive to improve the environment and did so without being forced by regulations. They were, in fact, *setting* the standards rather than complying with those set by others. In both situations, the companies gained a competitive advantage because of their actions, although Henkel's advantage was only temporary - until the other competitors could catch up. In the case of Body Shop, however, replication was not so easy; rather than just one new idea, an entire social/environmental company culture was created.

Industry self-regulation - and the 'proactive' response A third approach to improving the environment is an extension of market-based incentives and the productive response. Here, industry groups have recognised that it is definitely in their interest to determine the environmental standards - based on their own assessment of what is most economically viable - and to proactively reach environmental goals which exceed those established by regulatory agencies. Industry self-regulation also aims to set targets for *all* companies in the sector in order to ensure that some competitors will not be able to achieve lower costs by ignoring or creating pollution. This approach is also an alternative response to the command and control method. Industry groups have accepted that environmental issues are vital and cannot be avoided; therefore, it is better to take a proactive approach than to wait for unwanted bureaucratic standards to be imposed. Fundamentally, proactive self-regulation is based on the principle that the industries know best what is possible, what can be achieved at a reasonable cost, and the optimum way to focus technological efforts on environmental issues.

---

[4] Eutrophication is the process where a body of water becomes rich in dissolved nutrients (as phosphates) and becomes seasonally deficient in oxygen.

[5] This was a low-risk procedure as these natural ingredients were already being used for hair and skin care by people living where the ingredients originated.

Enlightened industry specialists firmly believe that *they* can develop a set of guidelines and actions which will produce much better overall results than anything proposed by governments, and that a 'level playing field' for everyone in the industry will foster compliance.

Three industries - automotive, chemical, and construction - have been proactive in taking self-regulatory measures. They have successfully led the way on environmental matters by developing their own technologies, programs and initiatives to protect and preserve the environment. These initiatives, which have tended to go beyond government regulations, have aimed to achieve two goals simultaneously: protecting the environment and remaining competitive.

The self-regulatory approach gained momentum after certain industries suffered significant problems with their cost structures by being over-regulated from an environmental point of view. For example, when California adopted its first air pollution law in 1972, the automotive sector was directly and seriously affected. The standards stipulated were not only very strict but, according to the manufacturers, 'beyond the technological and economic reach of the entire automobile industry.' It was painful, time-consuming and expensive for the automotive industry to prove that it was not technologically and economically feasible to achieve the established standards at that time. After a long process of confrontation and negotiations, more realistic targets and timetables to reach them were set. The sad part of this story is that so much energy went into fighting over the standards - instead of improving the environment. Bitter as the experience was, many industry experts concluded that it was absolutely *imperative* to adopt better environmental standards on a proactive basis - before they were imposed by those with an extremely limited understanding of how their actions would affect competition.

Self-regulation was the response to the German government's requirement to adopt air pollution control technologies. Inspired by Japanese firms which had reduced their nitrogen oxide and carbon monoxide emissions, Environment Minister Klaus Toffler adopted very stringent air pollution standards for CO_2 and NO_x. Initially, German companies had to purchase air pollution control products from Japan at rather high prices. Then, some German equipment manufacturers perceived that an opportunity, rather than a problem, was available. They developed cost-effective solutions that surpassed the standards - and thereby established the defacto standards for Europe. Five years after the Toffler standards were set, the Germans were selling and exporting the major market share of air pollution control technology in Europe. Moreover, it is the German companies that now have developed environmental standards which exceed governmental regulations.

Proactive self-regulation has occurred, to a large extent, on an industry basis. A good example is the voluntary program of the chemical industry developed during the 1980s under the name, 'Responsible Care (RC)'. Realising that every time a chemical company had an environmental accident, the entire chemical sector suffered from bad publicity and tighter governmental regulations, the chemical industry decided to start a program to prevent environmental damage. The program included periodic voluntary environmental auditing and a pledge of immediate technical assistance whenever an environmental accident affected a particular company. Today, the program is operating in 23 countries including both developed and developing countries. Signatories to the Responsible Care program commit themselves to managing their business according to the following ten principles:

* Recognise and respond to community concerns about chemicals and our operations.

* Develop and produce chemicals that can be manufactured, transported, used and disposed of safely.

* Make health, safety and environmental considerations a priority in our planning for all existing and new products and processes.

* Respond promptly to officials, employees, customers and the public with information on chemical-related health or environmental protective measures.

* Counsel customers on the safe use, transportation and disposal of chemical products.

* Operate our plants and facilities in a manner that protects the environment and health and safety of our employees and the public.

* Extend knowledge by conducting or supporting research on the health, safety and environmental effects of our products, processes and waste materials.

* Work with others to resolve problems created by past handling and disposal of waste materials.

* Participate with government and others in creating responsible laws, regulations and standards to safeguard the community, workplace and environment.

* Promote the principles and practices of Responsible Care by sharing information and offering assistance to others which produce, handle, use, transport or dispose of chemicals.

The RC program has a broad coverage. For example, the last point includes not only chemical companies but also the entire business system associated with chemicals. What the RC program does *not* do is establish exact standards for particular products and companies. The individual companies are responsible - according to their respective situations. Clearly, the chemical industry has received the greatest amount of environmental scrutiny, and its response has indeed been proactive. Many companies that have benchmarked successful approaches to handling environmental issues found excellent examples in this industry.

Sustainable development - and the 'visionary' response The term 'sustainable development', as applied to the environment and business, essentially addresses the critical interrelationship between economic growth, environmental quality and social welfare. A major objective is to understand the effect of increasing resource consumption by both developed and developing countries, as well as to introduce new technologies that will support and sustain a quality lifestyle without incurring severe environmental costs. According to WCED research[6], many new technological discoveries have created a modern society where life is better for some, but this 'progress' has also resulted in a rapidly deteriorating environment, destruction of the ozone layer, depletion of forests and lakes, excess waste and increasing poverty. Sustainable development seeks to determine a more acceptable tradeoff between

---

[6] World Commission on Environment and Development (WCED), *Our Common Future*, Oxford: Oxford University Press, 1987.

enhancing the lifestyle and environmental destruction. Making sustainable development really work will not be easy, yet the world has reached a point where we *must* seek ways to achieve it.

Industries that embrace the concept of *sustainable development* are using a 'visionary' response. That is, they specifically develop new technological processes, products and services that are congruent with the functioning of ecosystems and economic growth. The visionary response involves making radical changes in the way a company conducts its business. Thus far, only a few front-running companies are attempting to operate in this fashion. New concepts that are emerging as a result of industry efforts to understand and adopt the concept of sustainable development include:

* *Eco-efficiency* This approach enables companies to achieve economic and environmental targets simultaneously. One example is Xerox's Environmental Asset Management (EAM) program, whereby every aspect of a product's life cycle seeks to minimise the environmental impact and to provide savings through parts recovery, re-manufacturing and recycling. The key is to include all the elements of environmental quality at the outset of the design process. Thus, the design includes the product's *total* life cycle - not just its form, fit and functions, but its disassembly, re-manufacturing, reuse, recycling, and final disposal. As well, specifications and design must incorporate environmental requirements, which are then enforced by conducting mandatory reviews.

* *Design for environment* Many companies are reacting to society's strong feelings about conserving the environment. Instead of providing pollution control answers to environmental problems, they are adopting a systemic approach - assessing the entire environmental impact of their operations. In other words, the environment is taken into consideration when a product or service is designed. For example, the construction industry is considering joining forces with the telecommunications industry in order to avoid having additional conduits and installations for lighting systems. The goal is to control the lighting for a room from inside the telephone unit.

* *Zero pollution companies* Monsanto also used a visionary response in 1988 when it pledged to reduce the company's emission of 307 hazardous chemicals by 90% by the end of 1992, and to work towards the ultimate goal of zero emissions worldwide. From a base of 18.4 million pounds in 1987, Monsanto reduced its US emissions to 7.8 million in 1990, and 1.8 million by the end of 1992. The 90% emission reduction program marked a turning point in Monsanto's culture. Managers who had once questioned the wisdom of going beyond compliance with the law began talking about manufacturing plants of the future - plants with very few or no emissions. Monsanto found that self-imposed aggressive targets are realistic targets, and that these targets have created a new way of thinking in the company. Today, Monsanto employees believe that *they* are setting the environmental agenda for the future - based on *their vision* of what the environmental future should look like.

* *Distribution for environment* A leading European electronics manufacturing business, Design for Distribution, was one of the first to be certified for compliance with the UK environmental management standards BS 7750 (see Chapter 8). The company invented an effective distribution system for its products, which avoids potential environmental problems and improves the environmental performance of its suppliers.

Comparable to self-regulation by industry, some important groups are fostering the move to sustainable development. One example is the Paris-based International Chamber of Commerce (ICC), which has produced the ICC Business Charter for Sustainable Development containing 16 principles for protecting the environment and encouraging sustainable development. This document was developed by environmental managers from major companies and industry associations in Europe, and North and South America. The preliminary document was circulated to all ICC National Committees around the world for comments. The ICC Executive Board adopted the Charter in November 1990. The importance of the ICC charter is that the organisation includes over 1000 companies and associations around the world, including half of the Fortune 500 companies and others from countries such as Turkey, India, and Malaysia.

Sustainable development and taking a visionary response has also encouraged a new level of dialogue between people with environmental interests. For example, some leading-edge companies now realise that environmental pressure groups often have better information on certain topics than they do. The result is a changed relationship in many cases - one where the classic adversarial situation found in the command and control approach is being replaced by greatly improved information sharing due to a common interest in sustainable development.

Strengths and Weaknesses of the Business World's Response

The four approaches to improving the environment and the business response described above cannot be totally separated. Although they comprise four distinct methods for improving the environment, each one shares some elements with the others. Trying to determine which of these approaches is better for society in general and business in particular is not easy.

Command and control: strengths and weaknesses From an evolutionary viewpoint, command and control served as a starting point for raising consciousness - as well as one method where all companies in an industry could feel that they were being treated in a similar manner. This approach is still widely used - and necessarily so. Some believe that the proactive methods work better, with some asking why these methods were not attempted before the command and control era. The fundamental strengths and weaknesses of command and control and of the reactive responses to it are:

* Command and control is well known and understood by the various stakeholders. Its results are relatively certain and consistent. It is 'fair' in that the same standards are applied to multiple stakeholders. It is particularly effective when applied to a narrowly defined environmental problem which occurs in many places.

* International standards tend to be difficult to design and enforce. When the environmental problem is dynamic with multiple sources, it is difficult to apply rigid standards. Command and control is often applied when it is too late - a serious problem not only exists, there is no uniform agreement as to its consequences. Compliance can be prohibitively expensive. Usually the approach is inflexible, thereby excluding clever approaches that might be very cost effective. In some cases, command and control exposes competitors to grossly different compliance costs. Finally, the adversarial nature of the response to command and control means that compliance is a long drawn out battle - with excessive time and energy expended on non-productive activities.

Market-based incentives: strengths and weaknesses Market-based incentives often focus on a larger system. An example is a recent initiative by the Norwegian government to encourage Norwegian companies to invest in air pollution control in Eastern Europe - rather than only in Norway. The overall benefit is much higher for the same cost and, at the end of the day, everyone breathes the same air. The primary strengths and weaknesses of market-based incentives and the productive response are:

* This approach is flexible, it deals with the problem in the economic context understood by business, it can be cost effective, it allows those with particular environmental capabilities to use them more fully, it promotes technological development, and it tends to be especially effective when the pollutants are widely dispersed.

* The marketplace for this approach is not well developed, and it may function inefficiently and ineffectively. It is difficult to determine the cost/benefit relationships, making determination of incentives difficult. Some problems may be intractable for a market solution. It is difficult to police. The distribution of pollution permits may favour dirty industries over clean ones. It is less easy to understand - and to accept by those who do not believe in tradeoffs. It does not ensure a minimum level of improvement.

Industry self-regulation: strengths and weaknesses Industry self-regulation and the proactive response often provide the most effective way to improve environmental quality, since they continually examine a larger set of environmental problems. At some point, it is clearly less interesting to reduce some pollutant beyond its present stage - than to take on a new one. The primary strengths and weaknesses of this approach are:

* It is the most flexible approach, building on what an industry group sees as both important and attainable. It is cost effective by design. It tends to allow for an 'even playing field' for all the players in an industry. It supports innovative solutions - and continuous improvement.

* The complexity of the particular environmental problem may exceed the capacity of a single industry sector to solve it. It may be difficult to achieve consensus among individual companies in a particular industry. There will always be critics who see these actions as self-serving and perhaps as a subterfuge to avoid the real issues.

Sustainable development: strengths and weakness Sustainable development and the visionary response is clearly in an early stage of development. In fact, it is not yet possible to give a precise definition of sustainable development. But, the concept is heading in the right direction: we need economic development and we need to achieve it with the least cost to the environment and society. The strengths and weaknesses of this approach are:

* Sustainable development explicitly forces one to look at economic and social needs and environmental costs. It sets an ambitious objective that focuses attention on both the actions and the measures of effectiveness. It provides a means for organisations to communicate across industrial demarcations. It facilitates information being gathered from many sources.

* Sustainable development is still a fuzzy concept for most people - it is generalised rather than oriented toward environmental decision-making at the level of the firm. It might offer

opportunities for manipulation. It needs to have a more concrete and practical system of measurement.

Best practices Increasingly, informed and enlightened companies identify the environment as a means of developing competitive advantage. Previously, industry and business were not conceptually opposed to the idea of promoting environmental conservation, but when governments and pressure groups demanded actions that were seen as costly, difficult to accomplish or even threatening to a business activity's survival, the business sector tended to respond defensively. In more recent years, however, many companies have come to regard the environment as providing an opportunity to improve their business performance, increase customer perceived values, reduce costs and ultimately become more competitive.

Many companies that participated in the Business Council for Sustainable Development (BCSD)[7] at the Earth Summit in Rio de Janeiro claimed that incorporating environmental issues into business strategy is a 'win-win' situation. A number of examples prove the value of this approach. In Beaumont, Texas, a team at a Du Pont Chemical plant was able to reduce the ammonium sulphate emissions resulting from acrylonitrile manufacturing from 100 million to 40 million pounds per year, and slashed annual manufacturing costs by $1 million with no capital investment required. In Mehoopany, Pennsylvania, a Procter & Gamble plant manufacturing tissue, paper towels and diapers implemented a comprehensive program to minimise solid wastes, which resulted in an estimated savings of $25 million per year.

These companies understood clearly that it was in their interest - that is, they could become even more competitive by dedicating the resources and time needed to design products and services that were compatible with the environment, as well as leveraging from product/service innovations. This approach definitely worked for Volvo and Bosch. In 1972, they set up a joint venture to develop a device which would reduce the harmful effects of automobile exhaust emissions. This technology became the basis for the catalytic converter used by the automobile industry. In 1975, California was the only state requiring all automobiles being sold in the state to be equipped with this technology. Today, catalytic converters are mandatory in approximately 22 industrialised countries as well as a few developing countries. Volvo and Bosch have profited enormously by their visionary approach towards the environment.

Concluding Remarks

In this chapter, we focused on understanding how the environmental movement evolved and where the pressure points for change were derived. We have tried to show that business has responded to environmental pressures in a predictable manner, but its reaction has also been sporadic. As environmental concerns have evolved - into an ever more unified pattern, companies are increasingly realising that environmental issues are on the agenda, and will not decrease in intensity or quantity. Therefore, businesses must seize this opportunity to use new methods and tools for supporting more strategic, longer-term responses to these challenges and thereby develop a competitive advantage in their industry. That is, we come back to the

[7] The BCSD and WICE, The World Industry Council for the Environment merged in 1995 to form the World Business Council for Sustainable Development, WBCSD.

'environmental imperative': dominant businesses will set the environmental agenda, define excellent environmental performance, and be cost effective and competitive.

Environmental benchmarking is a process that companies in every industry can use to improve environmental performance while maintaining a strategic focus. The following chapter will outline the history and art of benchmarking, link them to improving environmental performance, and then introduce a process for environmental benchmarking that has been applied in the MIBE Project.

Chapter Three

BENCHMARKING ENVIRONMENTAL PERFORMANCE

Chapters 1 and 2 stated our fundamental objective for environmental benchmarking: how can firms incorporate environmental dimensions into 'best practices' that conserve the environment while increasing their competitive advantage over the long term? A firm may be sincerely interested in preventing and minimising all forms of pollution and securing cleaner air, better water and other measures of environmental improvement, yet at the same time it has to be concerned about being profitable. In fact, we argue that coupling environmental improvements with rational business decision-making will provide an even *better* impact all around. Business people generally make decisions within certain frameworks. Our intent is to demonstrate clearly how environmental decisions can and should be included in the best of these frameworks - rather than rely solely on social responsibility criteria for improving the environment. This chapter is devoted to answering the following questions:

* What *is* the art and practice of benchmarking? What have been its uses and payoffs, and why does it make sense for achieving superior environmental performance?

* How can benchmarking be linked to environmental improvement - and to other improvement programs such as total quality management (TQM) and health and safety (H&S)?

* Where should a company do environmental benchmarking? How are the results evaluated, and can benchmarking increase competitiveness?

* Where has benchmarking been used to improve environmental performance and competitiveness?

The Art and Practice of Benchmarking

There are many misconceptions about benchmarking, and it is essential to be definitive as to what environmental benchmarking is and what it is not.

Origins of benchmarking To some extent, benchmarking originated with the Japanese, who began over 30 years ago to initiate 'study trips' abroad to look at how others did things. They went to companies and they made careful observations, even taking photographs when allowed. Some say that the Japanese then 'copied', but in actuality they went beyond copying. They picked the best parts of everything and combined them to arrive at the 'super best.' The best was searched out in all spheres. Thus, the Toyota just-in-time manufacturing system had its roots in looking very carefully at American supermarkets, where there was no back room inventory and goods arrived as they were needed. But the 'looking' had to be done correctly. Detailed process analysis, hard work, adaptation and a philosophy of always asking 'why not' were key contributing factors. To a large extent this process defines today's 'benchmarking', i.e., smart copying based on the best of the best.

Xerox is clearly the best known firm for using benchmarking to great advantage, starting its benchmarking efforts in 1979. Xerox is an excellent example of a turnaround company, going from a market share of virtually 100 percent in the copier market, down to being nearly squeezed out, then coming back to become a major force in the industry again. At one point, the company determined that its competitors were selling copies for less than it cost Xerox to make them. This cost disadvantage simply had to be overcome. The Xerox cost structure was highly dependent on purchased components, which meant that the company not only had to improve internally, the entire supply chain needed to significantly enhance its performance. Moreover, better performance was necessary in more areas than cost reduction. The copiers themselves required substantial upgrading, with particular attention to reducing the need for so much field service support.

Xerox had many 'arrows in the quiver' as it sought ways to respond to these challenges, but one major activity was benchmarking. Xerox compared its processes with hundreds of other companies - and continues to do so today. An interesting question that arises is whether or not Xerox would be willing to compare its practices (i.e., benchmark) with any newcomers to the benchmarking game. The answer is undoubtedly 'yes,' with the possible exception of direct competitors - if for no other reason than Xerox would consider it a marketing opportunity.

There are at least two categories of companies that would seek to benchmark with Xerox: those that know what they are doing and those engaged in 'recreational benchmarking'. The latter would propose some poorly conceived comparison, such as looking at too broad an area - say for example, purchasing practices. In this case, the visitors might well get a talk from some public relations person, a nice lunch, a factory tour, and a one-hour question/answer session with a junior member of the purchasing department. If, on the other hand, the request focused on a very specific process, complete with flowcharts of a starting process plus the re-engineered process, along with a detailed set of questions and information on each member of the benchmarking team, plus an offer to show Xerox several things as a quid pro quo, then Xerox would be more inclined to think that these people might be in a position to provide an idea or two.

The practice of processes The most fundamental issue in benchmarking is that it must focus on processes. Processes are a series of stages that must be undertaken to carry out an activity. A process by definition can always be flowcharted, and conversely, if something cannot be flowcharted, it is not a process. For example, corporate culture cannot be flowcharted. Moreover, the culture of another company might be highly desirable, but trying to copy it is extraordinarily difficult. A culture just has too many dimensions. Of course, some of the dimensions are also processes, and *they* can be studied and *adapted* to another company's situation.

Similarly, a company's competencies or abilities - such as 3M's pollution prevention expertise, Proctor & Gamble's supply chain management, or another's corporate ability to champion the environment - are not processes. They are critical to the competitive performance of these companies - but they are not processes - and are not easily copied. There are many processes which support these competencies, and we will explicitly examine the linkages between processes and competencies later in this chapter. But, the key point here is that it is almost impossible to benchmark competencies - at least in terms of how we define benchmarking.

Practice is much more than measurement Benchmarking must not focus only on measurements. The belief that benchmarking merely involves comparing factories or other operating units in terms of their financial measures or environmental performance indicators is fundamentally incorrect. Benchmarking is concerned with processes, and the objective is to learn how to design and operate those processes in a much better way than is currently happening. That is, the emphasis must be on the practice of processes, with the measures as the result, i.e., they are the passengers not the drivers. The same set of rules applies to environmental measures. It is interesting, but not sufficient to compare two plants in terms of the parts per million of some toxic substance released into the environment. The much more important question concerns the practices of the particular processes that *cause or reduce* these emissions.

This is not to say that comparisons based on measures are worthless. Far from it. Knowing which companies are achieving superior emission reduction performance can lead to a series of detailed questions as to why they are successful. What processes support that level of performance? Who are the leaders in different industries and why? Who is achieving the best environmental performance? Who is ahead? The last two questions often bring many companies to study the chemical industry, even when they have nothing to do with chemicals. Chemical companies have been working hard on the environment for many years. There has been a great deal of attention focused on the environmental costs of chemical manufacturing, and successful chemical companies have learned a lot. In every case, the issues are less concerned with the actual results (passengers) than with processes (drivers).

We often ask managers if they compare performance in their various factories or other operating units. Most say yes, but they are much less definitive when asked whether the comparison works. There always seems to be what we call the 'yes but' answer regarding these comparisons:

* Yes, but my suppliers are not very environmentally proactive and knowledgeable.
* Yes, but I have many smaller uncommitted customers and the other has a few large committed ones.
* Yes, but I have to sell my products in another country - and they are not concerned and will not pay for greener products.
* Yes, but given the age of my plant, the waste I produce is much less.

There is still another problem with this type of comparison. They tend to be top down and they tend to focus on finding fault, which inevitably implies that the one being compared is somehow inferior. This impression leads to expressing reasons why this is not so, which is basically not productive.

Benchmarking needs to focus on processes and the practice of processes. Moreover, the work needs to be rigorous - not 'recreational.' Serious benchmarking requires that business process re-engineering (BPR) be included: that is, a process is selected and then carefully studied. It is examined with BPR techniques, including a detailed flowchart and process analysis which identifies each step in the process, quantity of resources consumed at each step in the process, or the amount of waste produced, or the energy consumed, and the number of people who are involved in the process, the steps which do (and do not) add value in terms of the customers' and other stakeholders' perceptions.

When this work is done carefully, potential improvements that should be implemented will become apparent. It is only at this point that serious benchmarking can be initiated, after the team knows how to look intelligently - how to examine other processes with a positive view toward changing their own. The examination needs to be done in the minutest detail; thus the need for the team to create a flowchart and process analysis for the particular process being studied. Only in this way can the team truly understand which details can be improved in their own operations and services.

The key here is to understand the means to achieving real improvement. There is a natural tendency for groups to defend their own way of doing things, a roadblock that needs to be overcome. Many times after a group has visited another factory, they carry on a lengthy conversation discussing why the other facility was not as good as theirs. We have heard these lines over and over.

* There is no way our customers would accept that mix of materials.
* They certainly don't do environmental audits very well.
* Did you see the amount of waste produced at the moulding stage?
* If the EPA caught us doing that, they would shut us down in a minute.
* What an expensive emissions tracking system they have in place. There is no way our management would allow us to justify that expense.

True benchmarking is able to overcome this 'not invented here' syndrome. One recommended approach is to have the benchmarking largely done by the owners of the process. That is, rather than someone from upper management putting people in a defensive position by telling them that they are X percent worse at something than someone else, the idea is to have people make the discoveries themselves. Moreover, because true benchmarking includes business process re-engineering as a pre-requisite to visiting other firms, the focus is on the details of the process. The team of process owners knows what to *look for* - how to note the details of the process, comparing them with their own process while searching for every aspect that is truly better. The goal is to make continuous improvement and to *always* assume that something of value can be learned. This approach also removes the concept of blame: the process owners can be motivated to improve their process without feeling guilty - just as a machine operator cannot be faulted for not having the newest machine.

Linking Benchmarking to Environmental Improvement

Improving the environment is a complex task, one that necessarily involves many people in the company. If the environment is only considered an issue that pertains to a few individuals - or even worse, as some kind of public relations activity - it will always be peripheral to the company's most important concerns. This is, in fact, the sad truth in far too many firms. Achieving breakthrough environmental performance - performance that yields a competitive advantage - requires mobilisation of resources, the commitment of virtually everyone in the company and a good sense of direction regarding how dealing with environmental issues can be improved.

At this point, we shall specifically define environmental benchmarking:

* *Environmental benchmarking* is a structured approach to rigorously examining and comparing, from an environmental perspective, the processes supporting different business activities. The objective of environmental benchmarking is to identify and assess the abilities and attitudes a company must have to excel in business and environmental performance simultaneously.

Identifying and assessing a company's abilities and attitudes includes finding out how such characteristics were developed and put into place. When benchmarking environmental performance, it is paramount to remember that we are not examining and comparing companies that have achieved good environmental performance in an inefficient and costly manner. Rather, our intention is to focus only on companies whose environmental performance is excellent while simultaneously remaining competitive. Moreover, our aim is to specifically seek companies that have increased their business competitiveness by improving their environmental performance.

For example, AT&T is a company that has reduced its business risks by adopting a proactive approach towards the environment. AT&T's corporate strategy acknowledges that in order to avoid future liabilities, the company must stay ahead of governmental regulations. Thus, the environmental strategy emphasises pollution prevention, waste minimisation and recyclability. Its mission includes bringing people together by making communications more useful to customers. Basic to this underlying mission is the company's objective to be profitable, and to be defined by its people and the values they share, particularly to encourage diversity in the workforce and a belief in teamwork.

Incorporating environmental thinking into ongoing improvement efforts Environmental issues need to be explicitly coupled with existing improvement efforts. Total quality management (TQM) is an excellent example. TQM is a management philosophy with four basic principles: continuously making an ongoing effort to improve business processes; doing things right the first time by recognising and eliminating problems before they occur; taking a systems approach to work by looking at each part of management or manufacturing as a system; and, most importantly for TQM, remembering that the customer is always right. TQM is being successfully applied in many firms, but not all firms succeed in making it work well for them. The ingredients for success seem to be good leadership, mobilisation of the entire workforce, a commitment to looking continuously for improvement, making periodic reviews, using a cross-functional approach, and team building. The four principles of TQM are also compatible with improving the environment. Several companies have redefined their approach to be 'Total Quality and Environmental Management' (TQEM). The focus on processes, empowered employees, continuous improvement and a customer focus are readily adapted to an environmental set of objectives.

Deming's 14 principles - applied to the environment An excellent extension of the TQM philosophy has been done by Philip Green in his revision of Deming's 14 principles of quality for application in an environmental context.[8] This philosophy is well worth studying; those who have already adopted Deming's set of ideas quickly see their value.

---

[8] Green, Philip E.J., "Environmental TQM", *Quality Progress*, May 1993.

1. Create constancy of purpose toward minimising detrimental environmental effects with a plan to remain competitive and stay in business.

2. Adopt the new philosophy. You are in a new environmental age. You can no longer live with accepted levels of contamination and environmental destruction.

3. Cease dependence on end-of-pipe treatment of waste streams. Require, instead, that you have statistical evidence that environmental protection measures are already built into the processes. Purchasing managers have a new job and must learn it.

4. End the practice of awarding business on the basis of the price tag. Instead, depend on meaningful measures of environmental protection, quality and price.

5. Find problems. It is management's job to work continuously on the system. It is not enough to wait for the environmental inspector to knock on your door.

6. Institute modern methods of training employees in environmental protection on the job.

7. Institute modern methods of supervision of production workers. The responsibility of the foreman must be changed from sheer numbers to environmental protection, quality and the elimination of waste.

8. Drive out fear, so that everyone can work effectively for the company and so that employees will alert management to environmental problems.

9. Break down barriers between departments. People in research, design, sales and production must work as a team to foresee problems in production that might be encountered and cause detrimental environmental effects.

10. Eliminate numerical goals, posters and slogans for the workforce that ask employees to protect the environment without providing methods to do so.

11. Eliminate work standards that prescribe numerical quotas with no regard to environmental protection.

12. Remove barriers that stand between the hourly worker and his or her right to pride of workmanship and his or her right to protect the environment.

13. Institute a vigorous program of education and training.

14. Create a structure in which top management will push employees every day to meet the first 13 points.

Clearly, this list emphasises the reality that good TQM practice has just the ingredients needed for improving environmental and business performance. Several leading-edge companies deliberately link the environment and TQM, adding the new set of objectives to the existing methodology and practice. The Global Environmental Management Initiative (GEMI)[9] of the

---

[9] See description of GEMI in Chapter 8, Environmental Standards and Management Tools.

US is generally credited as being the first organisation to marry environmental management and total quality management. TQEM maintains the four basic elements of TQM adjusted to specifically meet environmental requirements:

* *Catering to the customer* means all stakeholders in the environment such as external organisations - regulators, legislators, communities and national environmental groups.

* *Continuously improving* means seeking ways to make environmental performance and systems better.

* *Doing the job right the first time* means recognising and eliminating environmental problems before they occur, i.e., focusing employee attention on the causes of environmental problems instead of the symptoms.

* *Using a systems approach* means looking at each part of environmental management as a system, i.e., including all equipment and people across organisational boundaries to form teams representative of the value chain of either the service or products being sold.

Often the environmental agenda has even succeeded in rejuvenating a TQM program that had grown a bit tired. That is, the common objectives for TQM had been largely met, but then the new environmental goals were able to breathe new life into the program for the people involved. The point is that TQM is a known business process for continuous improvement. It has proven methods for determining direction, selecting projects, forming teams and evaluating results that work and work well! When this successful weapon is applied to environmental improvement, the outcome can be equally good. One caveat to achieving success is having environmental performance and TQEM objectives in line with corporate strategy and culture so that breakthrough improvements will occur.

AT&T has undergone a serious TQEM exercise which has provided excellent practical results. In AT&T's 1992 environmental report, it is made clear that TQEM helped the company to translate management's environmental policy into measurable progress: 'TQEM depends on management by fact, on data that enables us to pinpoint root causes of environmental problems and to identify the most effective ways to fix them. Policy deployment, a major component of TQEM, is the management system used to select, plan, execute and monitor our environmental objectives. Beginning with our commitment from the top, the goals are then cascaded through the Business Units for action by each process owner.' In AT&T's past and current commitment to TQEM principles, the company has also adopted the practice of environmental benchmarking and has shared this experience through participation in the GEMI primer on the subject.

There are some important similarities between the way TQM and benchmarking function, and these commonalties can be leveraged. But there are also some key differences. Environmental benchmarking and TQM both use teams, both circumvent the normal functional hierarchies/decision-making, and both focus on the owners of problems and processes as being in the best position to improve them. But, TQM is more encompassing (involving everyone) than benchmarking, and it tends to focus more on improving the status quo. Benchmarking fundamentally aims to 'stretch' for higher objectives, seek an outward view for comparison, select a few key processes for improvement, and use teams which cross existing organisational boundaries.

Health and safety (H&S) is a similar ongoing process for improvement that has been readily expanded to incorporate more proactive environmental issues. Chemical manufacturing companies are a good example of this change. EH&S brings the environment forward from the traditional compliance-based or health and safety-based perspective. EH&S allows environmental improvements to be coupled with ongoing effective programs for change. The venue for change is simply expanded - but in a somewhat logical direction. We say 'somewhat' because the direction will naturally still tend to focus on environmental issues that have an impact on the health and safety of the workers - but is expanded to include H&S product use issues, as well as H&S issues occurring in the community due to pollutants - it is expanded but does not focus on environmental issues that are of a more strategic/competitive character. We do not want to lose track of this distinction, because it underlines the reasons for incorporating benchmarking into environmental thinking.

Linking environmental improvement to benchmarking Many people in today's management circles talk about benchmarking. For most companies, the concept of *selectivity* for benchmarking is not as well developed as it needs to be. As we will see in Chapter 4, the MIBE environmental business transformation diagnostic *directly* addresses the issue of selectivity. We believe that benchmarking is far too important (and costly) to use as a general improvement process for everything. The same argument holds for another 'hot topic' that is closely related to benchmarking: business process re-engineering (BPR). According to Hamel and Champy,[10] BPR is 'the fundamental rethinking and radical redesign of business processes to achieve dramatic improvements in critical, contemporary measures of performance such as cost, quality, service and speed.' BPR again is being widely applied, but we believe it is important to target its application to critical processes rather than to apply it everywhere and anywhere. It is too easy to confuse activity with success and to live in a fools' paradise where people think that if a lot of action is going on, it must be good. On the contrary, we believe that it is absolutely essential to direct environmental benchmarking and business process re-engineering toward those critical processes which develop the long-run environmental competencies needed to respond to the marketplace in the best way.

Before going on to a more definitive description of the benchmarking process, we need to make an important key point. Our experience is that the best practise companies no longer believe in tradeoffs. That is, it is not a question of 'either/or'; the answer is always 'and'. In this case, no company should try to choose between TQEM, EH&S, Benchmarking, or Business Process Re-engineering. All these programs can and should operate in harmony. It is, however, essential that attempts at benchmarking not try to reinvent what can readily be observed, copied and improved.

We strongly believe that no one should get bogged down by definitions: it does not really matter what the efforts to systematically improve a company's environmental performance are called, and there are good ideas in each of the above-mentioned improvement programs that can and should be adopted. But, in any particular company, there will be good and poor practices. For those interested in improving environmental performance, it is critical to tie their efforts to the winners. They will consist of whatever improvement programs are effectively operating in the company. Moreover, one's own company is only the *first* place to compare practices. The most efficient benchmarks are external - that is, using the best in the world!

---

[10] Hammer, Michael; Champy, James, *Re-engineering the Corporation: A Manifesto for Business Revolution*, New York, NY: Harper Collins, 1993.

The environmental benchmarking team The benchmarking team should be largely made up of process owners, and especially valuable are those who cross functional boundaries - as they will have access to the processes which are most important for environmental benchmarking. The team often also includes one or more individuals with benchmarking experience as they can help the others adopt the right attitudes and methods for achieving good results more quickly. As well, the corporate level may need to be involved, since environmental changes typically affect more than one function within the organisation.

At Baxter, each manufacturing facility must compare its performance with a set of 46 world-class benchmarks. In 1990, Baxter's Chairman and CEO decided that Baxter facilities worldwide would aim for 'state-of-the-art' (SOA) environmental performance goals by 1996. Achieving SOA was pushed down to the facility and unit level by making it part of the evaluation system for plant managers and their employees. This move mobilised Baxter's facilities worldwide into re-engineering their processes and practices as they prepared to benchmark their progress within the organisation. Today, Baxter is well on its way toward reaching its goals.

Breakout thinking True benchmarking does not allow being satisfied with a 10 percent improvement; rather, there should be at least a 50 percent improvement. At Baxter's Renal Division based in Castlebar and Swinford, Ireland, improvements of around 60 percent were necessary to meet SOA criteria for reductions in freon emissions and landfill usage. Both targets were met within a three-year period, the latter objective in a single year. This was accomplished by setting up a steering committee and several teams of employees from all levels of the organisation. In this case, the target was intentionally high in order to support the development of innovative solutions.

Figure 3.1 illustrates the values of environmental benchmarking. It is essential that the environmental targets be set high enough. Aiming for 50 percent improvement is a good benchmarking goal (so is the SOA approach). In fact, it helps to set priorities, since the attention is focused on areas with more possibilities for accomplishment. In general, the more organisational boundaries involved, the greater the potential for improvement. Another necessary ingredient for environmental benchmarking is overcoming disbelief; people must be convinced that conserving the environment *and* being cost effective are both indispensable for achieving excellent environmental practices. In fact, it is being able to reject the traditional 'tradeoff' thinking that separates excellent practice from the 'also rans.'

Accountability and culture change are the other values found in environmental benchmarking. Environmental performance and improvement need to be continually monitored, and the best ideas are always discovered when looking outward - by expanding horizons, making connections and collaborating with other environmental stakeholders. It is better to view environmental action groups and international organisations as potential sources of information and opinion shapers rather than as enemies and adversaries.

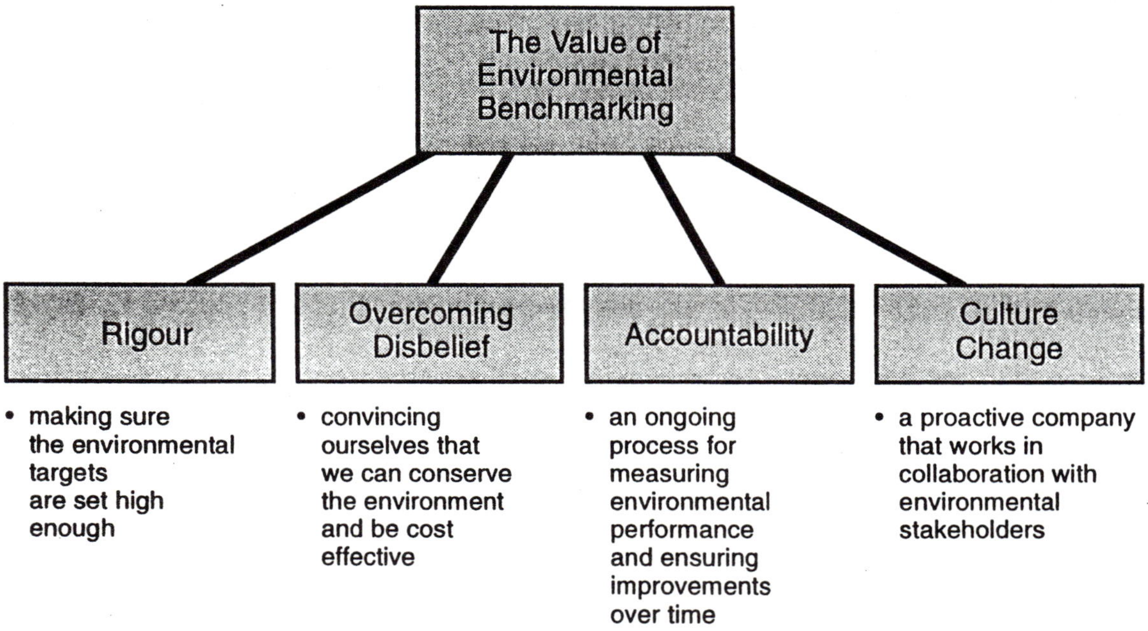

Source: Johnson & Johnson (adapted)

Figure 3.1 Values of environmental benchmarking

Where Should a Company Benchmark?

A key attribute of environmental benchmarking is to look beyond obvious comparisons. Initially, it is best to focus benchmarking efforts internally. That is, one should start by looking at a particular process (with business process re-engineering), then make comparisons with sister plants (to learn *how to look*), then try to make comparisons with competitors and the best in the industry (as far as possible). But, finally, the really important comparisons are those made *outside* the industry - for that is where the breakout ideas come from.

Scope of environmental benchmarking The scope of environmental benchmarking also needs to encompass all areas of the firm's activities. Improvements are not limited to manufacturing processes. Figure 3.2 shows the breadth of *internal inquiry* that is required. No part of the company should be considered immune from environmental improvement (and benchmarking) The accounting and auditing group must report on more than financial performance. As this area is the most familiar with reporting company conditions to outside bodies, it is the logical place to house environmental measurement and reporting systems. A growing number of the leading firms today are issuing a separate environmental performance report, but this task has usually been assigned to the communications and public relations departments. Thus, key environmental benchmarking decisions are: finding the best way to design environmental reporting systems; determining which measures to collect; systematically using the systems to support improved environmental practices; and learning not only how to comply with existing regulations, but to anticipate and perhaps even *drive* new environmental reporting practices.

Financial	Administrative	Manufacturing/ Services	Marketing	Technical
• Environmental Measurement Systems • Environmental Reporting Systems	• Legal Dept. • Employee Surveys • Environmental Communication System	• Total Life-Cycle Analysis • Adopting the 4R's: Reduce, Reuse, Recycle, Re-engineer • Design of Sustainable Environmental Systems	• Customer Satisfaction • Proactive Environmental Compliance	• New Material Alternatives • Technology Development • Technical Environmental Guidelines for Plant Managers

Source: Johnson & Johnson (adapted)

Figure 3.2 Internal scope of environmental benchmarking

A benchmarking system has been devised by SustainAbility Limited and published by the United Nations Environment Programme Industry And Environment (UNEP),[11] suggesting that there is still plenty of room for improvement in the environmental reporting practices in most companies. The report identifies five stages in corporate environmental reporting, as seen in Figure 3.3, and it ranks 100 companies that have adopted environmental procedures for these five stages. As well, 50 reporting ingredients (see Appendix A) are presented as 'a set of basic building blocks which companies can use to construct their reports according to their own priorities.' Companies that use these criteria as a guide are participating in a form of environmental benchmarking, since the environmental reporting practices of industry leaders were the major influence in the development of the criteria.

Referring back to Figure 3.2, the section dealing with manufacturing/services focuses again on those activities that will redefine environmental performance - rather than simply fulfil regulations. Total life-cycle assessment is a part of the environmental lexicon: it *must* be the common way of thinking in the company if the objective is to excel. Similarly, the '3Rs' should drive the way in which products and services are designed. The same is true for sustainable environmental systems: this needs to be ingrained in the way a company *thinks* about providing goods and services to its customers. In every area, there are a multiplicity of processes that can be studied and improved. It is critical that manufacturing and service people be charged not only with continually improving every process but, in fact, with consciously looking for breakout ideas.

---

[11] SustainAbility and UNEP, *Technical Report No. 24, Company Environmental Reporting, A Measure of the Progress of Business & Industry Towards Sustainable Development*, UNEP, ISBN 92-807-1413-9, 1994.

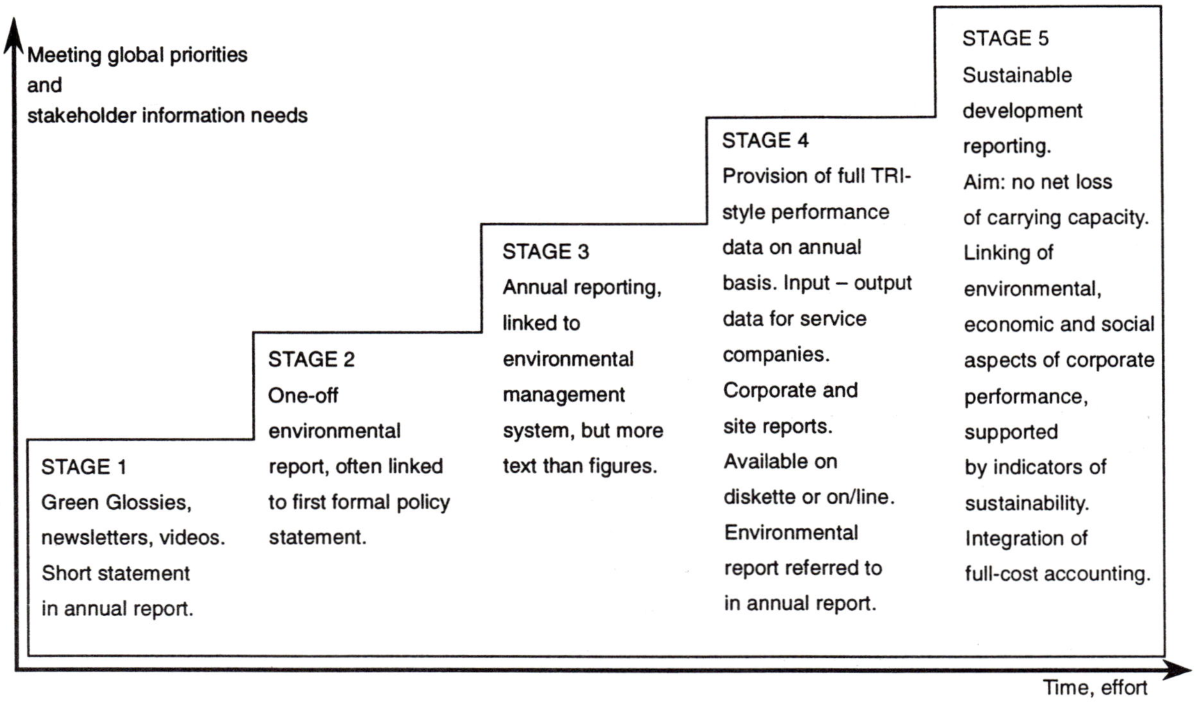

Figure 3.3 UNEP's five stages in corporate environmental reporting

Marketing and sales are similarly decisive in developing dominant environmental practices. A firm *must* know (and anticipate) what environmental issues are most important to its customers. If the company can be proactive on this front, it can move definitively away from commodity/price discussions with its customers into more interesting and mutually rewarding value/marketplace discussions.

Finally, Figure 3.2 focuses on the technical area of company operations. Here, the focus on new product and process development has to encompass the environment not as a constraint, but as a goal. The most important environmental variables are in fact determined in the design stage, which is where constraints on future improvements are built in. The technical group needs to adopt a very proactive stance on the environment, where superior performance is seen as a way to change the rules of the competitive game and lead the company toward product and service offerings with more sustainable growth.

The scope of environmental benchmarking can also be usefully examined along the business chain. Figure 3.4, the MIBE environmental wheel, illustrates this approach for both manufacturing and service companies. This wheel represents a conceptual environmental business approach for the 1990s and beyond. Between 1970 and 1988, most environmental policies, laws and regulations focused mainly on reducing pollution. For this reason, most countries throughout the world currently have more legal instruments in place to curb pollution than to conserve nature and manage the environment. When considering the environment from industry's point of view, it is important to include not only the contaminating emissions that an industrial operation might generate but also all aspects of the business system in which a company operates.

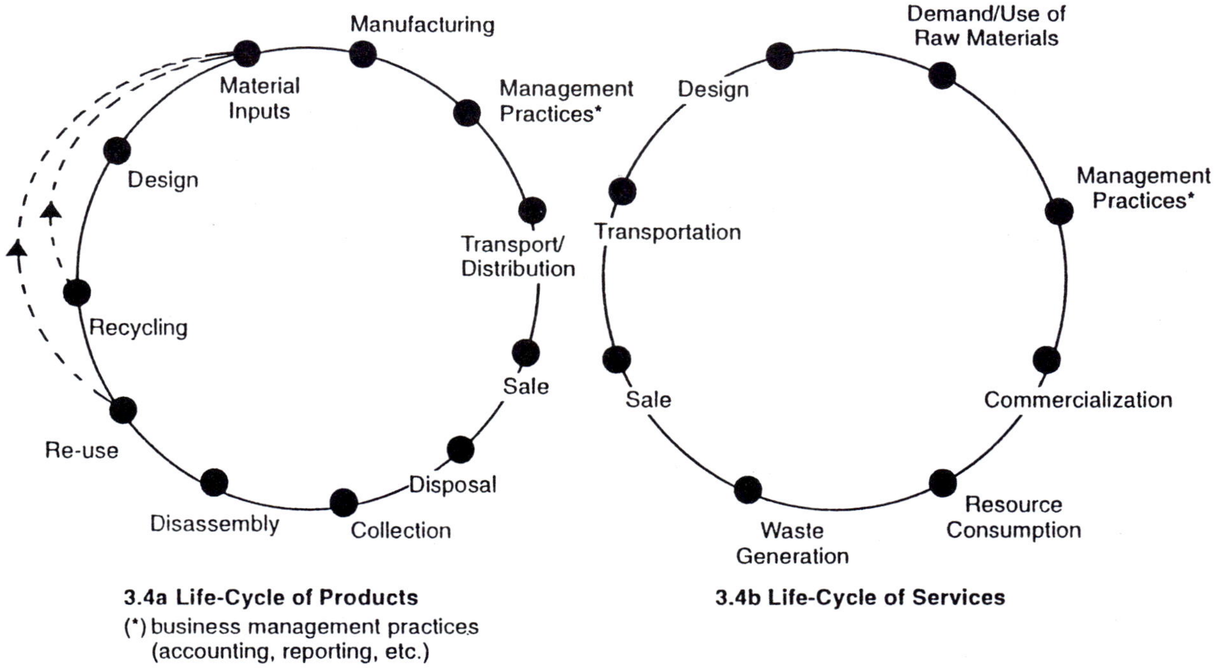

Figure 3.4 The MIBE environmental wheel

The flow of goods and services, from suppliers of suppliers to customers of customers, needs to be seen in terms of the environmental attributes effected at each stage - as well as those which cumulate in the chain. Breakout performance is often associated with taking a much wider perspective on environmental issues. The firm that clearly understands environmental performance as perceived by customers of customers - as well as understanding the competencies of suppliers and *their* suppliers - should be able to design superior solutions.

It is apparent in Figure 3.4a that by controlling pollution emissions, a company is only dealing with one part of its environmental impact. There are many cases where a company could have a 'clean' manufacturing process that generates very little or no contaminating emissions but demands scarce raw materials or generates products that are difficult to discard. The production of construction materials made of PVCs (Polyvinyl Chorides) is a good example. It involves a relatively clean process; however, disposing of these materials at their end-of-life is a different matter. Some countries in Europe, such as Switzerland and Germany, have regulated the incineration of PVCs - due to the creation of dioxins which are concentrated in the ashes of the incineration process.

In Figure 3.4b, it is possible to observe the business system as applied to service companies. Similar to the situation in the manufacturing sector, a service company must include environmental concepts in all stages of its activities. Many companies in the leisure sector are promoting a new green service called 'eco-tourism', which is compatible with nature and promotes the conservation of natural resources. It means diverting traditional local tourism to special places with apparently favourable environmental characteristics. For example, European travel companies are frequently advertising eco-tourism activities in the

Red Sea - that is, snorkelling in marine parks where fragile marine ecosystems will not be disturbed or destroyed. However, by encouraging people to travel long distances by aeroplane to the eco-tourism site, some of these companies will be increasing the overall environmental impact of tourism on nature!

One important corollary derived from carefully examining the MIBE environmental wheel is: to generate activities that are compatible with the environment and sustainable over time, society needs to have more precise information on *what is good for the environment.* Unfortunately, as we mentioned earlier, people are still in the early stages of collectively understanding how ecosystems function and what are environmentally sound and sustainable manufacturing processes, technologies and services. A more realistic approach towards these complex issues is to seek a path towards environmental excellence rather than absolute answers on how to develop perfect environmental solutions. Environmental benchmarking is the surest road heading in that direction.

Using Benchmarking to Improve Environmental Performance

By now it should be clear that we see environmental benchmarking as a means to make *significant* improvements in environmental performance - as a way to achieve competitive dominance. But, for most companies there will be an evolutionary process, based on a changing set of targets.

Approaches to environmental benchmarking As in traditional benchmarking studies, there are several different approaches to environmental benchmarking which companies have used. Unfortunately, the environmental performance of far too many companies is largely legislation and compliance driven, and typically is the last piece of 'business' on the agenda of the board of directors' meeting. As well, the environment is viewed as a 'cost' by most companies rather than the 'opportunity' that environmental leaders, entrepreneurs, stakeholders and educators espouse.

Benchmarking to ensure compliance with regulations is clearly a widely-used goal for environmental benchmarking. Most organisations have to comply with one regulation or another. Organisations need to have the appropriate systems to fulfil these requirements, and it is useful to find out how other companies do it, in their search for 'best practices'. Some firms hire consulting companies that have helped others comply with the regulations.

Another objective or approach to environmental benchmarking is to increase the cost efficiency of manufacturing processes. These costs may be traditional costs or incorporate environmental cost savings concepts, but the driver for improvement is *cost.* In environmental terms, it means that cost savings is synonymous with minimising waste. Organisations that have adopted stringent and successful environmental waste minimisation programs are numerous: 3M, S.C. Johnson & Son, Veryfine Products, etc. The companies which really do achieve a level of excellence in environmental cost effectiveness can reap competitive benefits. As of 1994, 3M's Pollution Prevention Pays program (created in 1975) resulted in 4,163 projects worldwide, 1.3 billion pounds of pollution prevented from affecting air, water and land, and has garnered over $710 million in savings. Similarly, National Westminster Bank in the UK registered energy savings of $15.2 million across its worldwide portfolio of 4,300 properties.

Still another approach to environmental benchmarking is developing environmental product/service attributes as a definitive competitive advantage. This approach is implied in the prior classifications as well, but here the emphasis is not only on compliance and excellence in reducing costs, but also on the marketplace. The company uses benchmarking to develop definitively superior goals, products and services in terms of environmental attributes. Benchmarking to support this goal requires the mobilisation of a high-profile team of managers, and the full commitment and involvement of the top level decision-makers in the organisation.

A final approach to environmental benchmarking starts with the conservation of the environment and sustainable development. Governments, environmental organisations and industry associations are still trying to define what determines a sustainable product, service or business. Thus, it is inevitably a difficult goal to achieve at the company level. Sustainable growth may mean conducting benchmarking studies on how to stop the production of specific products while maintaining the social stability of the community where the business is operating. The organisations that are actively participating in these debates are the most likely ones to establish benchmarks in the future.

The Thirteen Steps of the Environmental Benchmarking Process

The benchmarking process that was implemented for the MIBE Project at IMD consists of 13 steps that can be grouped into four phases of activity. These steps and the phases are shown in Figure 3.5. Why 13 steps? There is no magic number of steps to follow when it comes to developing or conducting a benchmarking process. Rather, it is being committed to the basic philosophy of the procedure that is important.

As mentioned in the previous chapter, benchmarking basically means making comparisons, and then adopting best practices wherever they are found - always looking for 'the best of the best'. The philosophy embraces the notion of striving for continuous improvement by learning from others. After thoroughly assessing one's own processes or management systems, the next step in benchmarking is to learn from other companies by studying both their successes and their failures.

The steps presented in Figure 3.5 can be used by companies as a basic benchmarking map or guide. The four phases are separated in order to focus on the fundamentally different activities. The first decision is to carefully select which processes need to be benchmarked. Next, some internal homework must be done - to discover how to look; then the actual benchmarking can begin. Finally, the process has to start all over again!

Phase one is referred to as the *Discovery Phase*. It is in this phase that the Business Transformation Diagnostic developed with the Manufacturing 2000 project at IMD is utilised. This phase, which is separated into two parts, addresses the question of *what* to benchmark - not how to benchmark. The first part deals with clearly articulating the environmental challenges facing the company - and what it wishes to do in response. The second part focuses on developing the appropriate responses to these challenges, and then identifying the competencies necessary to make those responses.

Phase two of the diagnostic is where the team concentrates on assessing the internal systems and key processes that are to be benchmarked. It is also the phase when the team decides which are the best organisations to use for benchmarking. This phase is referred to as the *Awareness Phase*.

Phase 1	1	Survey top managers, re. strategic intent, customer expectations, core products and services, and the resultant set of core competencies that are required for the **future**
	2	Determine capabilities, practices, processes, and resources for achieving key competencies
	3	Identify the processes that will **enable** these competencies
Phase 2	4	Select the benchmarking team
	5	Apply business process re-engineering
	6	Decide **where** to benchmark
	7	Do some homework — and **keep** doing it
	8	Select benchmarking companies
Phase 3	9	Make the visits
	10	Measure the gaps between observations and **your** practices
	11	Develop the action plan
	12	Report the results
Phase 4	13	Figure out what is next

Figure 3.5 Thirteen steps to environmental benchmarking

Action transpires in phase three. The benchmarking visits are conducted and the team analyses the differences between their company's practices, processes and systems and those of the best-practice companies targeted in phase two. The ultimate goal of this phase is to develop and implement the action plan for improving the targeted processes.

Phase four is where a company realises that benchmarking is one method to ensure *Continuous Improvement.* In regard to environmental practices, this concept is especially pertinent. The way in which business, society and governments interact with the environment is continuously changing, as does the definition of a green product, service or organisation.

An Integrative Example at Sulzer Technology Corporation

Following chapters The following four chapters give a more detailed explanation of the four phases of the 13-step environmental benchmarking approach. To illustrate the steps, we will describe the journey of Sulzer Technology Corporation, a Swiss-based technology corporation, whose worldwide operations account for an annual turnover of $5 billion. Sulzer's business policy is to utilise technology to strengthen the economy and improve the environment. As a company sponsor of the MIBE Project, members of Sulzer's Corporate Planning and Acquisitions group decided to embark on an environmental benchmarking process. The purpose of the study was twofold: first, to develop a skill in environmental benchmarking that the corporate group could then transfer throughout Sulzer's strategic business units; secondly, Sulzer wanted to use this process to develop a better understanding of the most critical impact Sulzer's businesses have on the environment. At the conclusion of each of the next three chapters, we will give a detailed description of Sulzer's experience as the company implemented the four phases.

Chapter Four

PHASE 1: DISCOVERY
What are the Company's Environmental Challenges?

Chapter 4 is devoted to the 'Discovery Phase' of environmental benchmarking. The goal is to understand, in concrete terms, what environmental challenges the company faces and the detailed responses that it must make to meet those challenges. This chapter will provide answers to the following five questions:

* What is the environmental business transformation diagnostic?

* How is a meaningful environmental strategy established?

* How is the detailed infrastructure matched to the strategy?

* How is facilitation of the diagnostic process supported?

* How does this work in the context of the Sulzer example?

The Environmental Business Transformation Diagnostic

Direction setting The key environmental challenges facing a company can be determined by using the Environmental Business Transformation Diagnostic (shown in Figure 4.1). The upper half of the diagnostic focuses on assessing the current set of internal and external challenges for improving the organisation's environmental performance. The environmental expectations of the customers and other stakeholders are assessed. This half of the diagnostic facilitates the process of reviewing what is happening both outside and inside the organisation. What are the pressures for change, what are the responses to these pressures, and how do they fit into an overall strategic approach to the environment? This analysis leads to a clearer understanding of what changes are necessary for the environment and what the company needs to do to address related problems and potential opportunities - including possibilities for developing a cross-functional environmental consensus. The most important link in this half of the diagnostic is ensuring that the strategic intent and the strategic response fit into an overall set of corporate goals and targets for corporate environmental performance.

Infrastructure The bottom half of the diagnostic focuses on the detailed infrastructure and the change agenda, i.e., the internal processes and system changes that are necessary to carry out the strategic responses that will address existing and future environmental pressures and challenges. More simply, it addresses what the organisation *physically* needs to change. Successfully implementing the results in the bottom half of the diagnostic depends heavily on the organisation's ability and capacity to learn - on its willingness to improve through the benchmarking process.

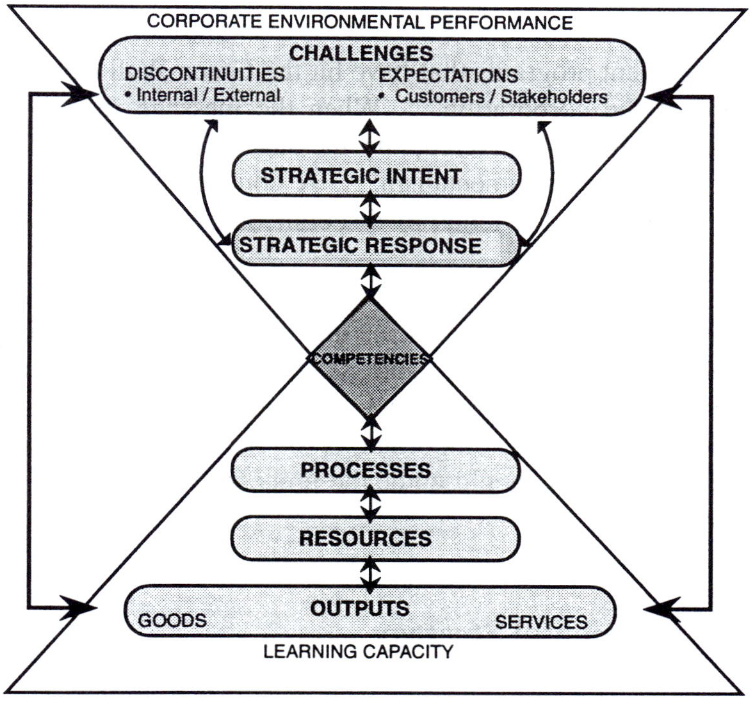

Figure 4.1[12] *The environmental business transformation diagnostic*

The fulcrum between the two halves of the diagnostic indicates the competencies required to make the appropriate responses to existing and future challenges. Identifying competencies is critical to the success of the benchmarking process. At this point, the team will use the diagnostic to move from the helicopter view of the upper half to a micro view of a few selected competencies and their underlying processes and resources. These selected competencies must be the specific ones that will enable the organisation to achieve a competitive advantage through improving environmental performance.

If the diagnostic has been applied correctly, a 'red thread' linking each of the components of the tool together can be identified. For each process that is to be benchmarked, every manager in the company should be able to follow a logical sequence back to the root environmental challenges/opportunities, via the desired response to these challenges and the competency required to *make* that response.

Cultural change It is essential at this point to ensure that the competencies identified fit the business culture and strategy of the organisation. Dealing with environmental 'language' is still a relatively new concept for most business managers. As indicated in Chapter 3, the 'E' in Environment has become prominent and has been integrated into existing functions - such as EH&S or TQEM principles. Nevertheless, environmental terms are often still difficult for many business managers to understand unless they are linked to traditional business terminology.

---

[12] Adopted from Vollmann, *The Transformation Imperative*, Boston, MA: Harvard Business School Press, 1996.

As stated by a Director of the Environmental Health and Safety Management Practice at Arthur D. Little:[13]

> Environmental management programs that have hit the Green Wall have failed to fit in with the business culture of the corporation. ... When the message regarding the benefits of environmental management are delivered, the message is often unintelligible to business managers because it is in environmentalism. ... A company has a business culture that speaks a business language and it tends to discount or exclude those who do not share similar attributes.

Clearly, then, it is essential that the competencies identified for benchmarking fit comfortably with the organisation's business culture and strategy. Success is more likely if the environmental business transformation diagnostic is facilitated by managers who understand environmental issues. The *journey* is as important as the *destination* - that is, individuals who share the same culture and strategy should guide the process of determining the environmental benchmarking objectives.

Developing the Environmental Strategy

If an effective environmental action plan is to be developed, the people involved in the change must fully understand how the environment is linked to traditional business goals, and how it can help the business become more efficient and competitive. For example, a serious concern for many businesses is how they utilise energy, i.e., a paper producer would like to cut the amount of energy consumption per unit of output. The main objective is to be more cost effective; however, by achieving this business goal, the company can also reduce its impact on the environment - since it will be producing fewer kilograms of carbon dioxide per unit of output. This kind of action plan, one that fulfils both business and environmental goals, represents a win-win situation.

Seeing these opportunities presents a significant challenge; companies are only in the early stages of understanding how re-organising their business with an environmental focus can provide competitive advantage. The situation is somewhat exacerbated by the compliance-based mentality that existed in the past. Also, the belief that protecting the environment is an expense rather than an investment with a payback has inhibited many businesses from making significant changes. Thus, for environmental benchmarking to be effective, it is vital to identify competencies that clearly correspond with the organisation's business strategy.

Competitive advantage The fundamental objective of the diagnostic is to focus on developing the skills and processes that will give the organisation an 'environmental/business' competitive advantage. The diagnostic is used to weave the 'red thread' that integrates the activities for the remainder of the benchmarking process. The result is a benchmarking study based on management's consensus on the environmental challenges facing the organisation and their proposed solutions.

---

[13] Shelton, Robert D. , "Hitting The Green Wall: Why Corporate Programs Get Stalled", *Corporate Environmental Strategy, The Journal of Environmental Leadership*, Autumn 1994.

Discontinuities The strategy development process begins by identifying the internal and external environmental discontinuities. Discontinuities are defined as the environmental forces or significant changes both inside and outside the organisation that could potentially improve the company's environmental performance. These forces or changes may already exist or may only be anticipated at some later date. Being able to project explicit future developments is extremely valuable, as the environmental business revolution is relatively new and constantly changing.

Examples of environmentally-related *internal discontinuities* identified in environmental benchmarking sessions at IMD include:

* Company A found that its organisation structure, culture and compensation/reward system was limiting its ability to meet environmental mandates. For example, good environmental performance was not one of the criteria used to promote people upward through the organisation.

* Company B discovered that the lack of structured information-sharing resulted in misinterpreting customers' needs and economic feedback from pressure groups.

The internal discontinuities most commonly cited are organisational in nature, i.e., structural and cultural issues that act as barriers to making environmental progress. Some examples of environmentally-related *external discontinuities* include:

* Company X had to deal with increasing local regulations that differed from company-wide mandates. Finding the investment funds to meet local regulations was becoming difficult.

* Company Y discovered a serious conflict between environmental needs and customers' real behaviour, i.e., customers were demanding products with a better environmentally-oriented performance, but they refused to pay higher prices to have improved products.

The most frequently identified external discontinuities are legislative pressures and cost pressures. Most organisations are wary of existing or pending legislation. As legislation varies from one country to another, these discontinuities are especially difficult for global organisations that are decentralised yet operate under one environmental policy.

Expectations Expectations relate to two separate groups: customers and other stakeholders. Other stakeholders include employees, shareholders, employee unions, environmental pressure groups, governmental agencies, the media, etc. Expectations are defined as being whatever environmental aspirations a company's intermediate customers, final consumers, company stakeholders and all others with a particular interest want satisfied. They resemble discontinuities as they also present a *challenge* to the organisation.

Customers are given special priority in this instance, because meeting or exceeding their environmental expectations is presumed to relate directly to creating competitive advantage. Companies may have several levels of customers: distributors, retailers and the final consumers of their products and services. The ability to focus on and listen to their customers is a key success factor for most businesses. This holds true not only for traditional customer demands but also where the environment is concerned.

Some examples of customer expectations identified in the environmental benchmarking exercises include:

* Company C was expected to minimise the post-consumer environmental impact of all its products and services.

* Company D's suppliers expected assistance with handling environmental issues related to their businesses.

Other stakeholder groups tend to gather information and develop criteria, metrics and assessment systems by which an organisation's environmental performance is measured. Achieving best environmental practice includes maintaining a dialogue with these groups as part of the information gathering, planning and assessment process - they are proactive players who have influence. Seeking the support of these stakeholders and understanding their aspirations will enable companies to be more proactive and influential in legislative decisions and setting standards that other companies will have to meet. Other stakeholder group expectations frequently identified include:

* Company V was expected to generate minimal or no emissions into the air, water and soil.

* Company W was asked to provide credible information about the environmental impact of its products and services.

Strategic intent Strategic intent can be defined as the company's environmental mission statement. It needs to reflect a strategy toward the environment that also coincides with its overall business strategy. The environmental strategic intent should have some longer-range 'stretch' objectives that strive and aspire toward even greater goals, that would result in a true competitive edge. More basically, the company's environmental mission statement needs to be operational, so that all levels of the organisation know how to make environmental decisions related to their daily tasks. To be precise, the environmental mission statement needs to be practical and realistic - not a mere public relations statement that causes internal confusion. An operationally viable mission statement may include, but is not limited to: environmental targets with deadlines and goals; a description of how the organisation plans to meet these targets and goals; explicit statements regarding employees and customers; policy with respect to legislative compliance; an outline of responsibility; and recognition or rewards for environmental performance. Bristol-Myers Squibb has produced a good example of an operational environmental mission statement (see Figure 4.2).

Bristol-Myers Squibb has further explained its overall environmental intent in four documents: The Bristol-Myers Squibb Pledge, Purchasing Guidelines, Environmental Guidelines for Package Development, and Product Life Cycle - Building Competitive Advantage brochure. Three of these documents can be found in Appendix B, they are concise in length (from one to one-and-a-half pages long) and can be applied across functional boundaries.

It is the policy of Bristol-Myers Squibb Company to protect the health, safety and quality of life of its employees and the public, and to exercise responsible stewardship of natural resources that may be impacted by company activities. To accomplish this, the company is committed to maintaining programs and procedures for the environmentally responsible management of:

Facilities
Materials
Production Processes
Products and Packaging
Transportation and Distribution
Waste and Its Minimization
Energy
General Business Operations
Contracted Goods and Services

Each facility shall support this policy in accordance with prevailing government regulations and the guidelines set forth in the Corporate Environmental Manual. Management at each location will document its programs in the form of written environmental procedures.

The president of each division shall develop and implement a plan to address the organizational and human resource needs required by this policy.

The Technical Evaluation and Service Department is responsible for conducting periodic evaluations of company operations and compliance with this policy and for reporting the results of such evaluations to corporate management and to the appropriate group, division and subsidiary presidents.

All questions shall be referred to the vice president of Environmental Affairs, Occupational Health and Safety.

Figure 4.2 Bristol-Myers Squibb corporate environmental protection policy[14]

Strategic responses Strategic responses are the specific actions taken to address environmental discontinuities and expectations, while fulfilling the organisation's strategic intent. Determining a strategic response is the culmination point in developing an environmental strategy - in meeting the challenges. Once the overriding objective is clear, the means of achieving it can be planned. We offer two examples of environmentally-related strategic responses:

[14] Excerpt from "Remarks to Bristol-Myers Squibb Environmental Conference", Richard L. Gelb, Chairman of the Board and Chief Executive Officer, Chantilly, Virginia, September 26, 1991.

* Company F needs to change both the content and the amount of environmental training and education for management at all levels.

* Company G needs to respond by developing a global environmental information network with regard to legislation, the position of green groups, market opinions and available environmental measurement tools.

Common environmental strategic responses Observations made while administering the business transformation diagnostic in IMD's MIBE Project and several of IMD's executive training courses resulted in ten groupings of environmentally-related strategic responses being commonly identified. The following is a descriptive interpretation of each one.

* *Leadership: Become the environmental leader in the industry - on a two-dimensional level* The first is by making a proactive response to environmental problems before a response is mandated by governmental agencies or public opinion. The second dimension is for personal fulfilment - through leading the organisation by example, taking action and meaningful decisions.

* *Information: improve the means for accessing information* There is a clear consensus that having better sources and a greater use of information technology would significantly improve communications with environmental functions outside the organisation. Varying aspects of environmental management, including market information, legislative information, and internal performance, need to be communicated through the right channels. Those who keep informed can stay ahead of the competition.

* *New initiatives: become a leading-edge company* Being committed to developing new business based on important innovations in the environmental design of products and services will help firms to reach the leading-edge level. Such initiatives include partnerships with customers and recyclers to develop environmental competencies in the overall supply chain.

* *Market research: exceed customer expectations* Exceed with regard to the environmental performance of products, services and throughout the organisation. To accomplish this goal, a new type of market research is necessary in many instances. Customers are unsure of what they want, what they are willing to pay, and whether they are willing to alter their respective lifestyles in order to accept more environmentally-friendly products and services.

* *Cost: develop environmental accounting systems* Environmental accounting systems that will incorporate environmental costs into decision-making more effectively. Companies also find it beneficial to develop improved cost/benefit analysis in order to have better decision-making for environmental projects. Some organisations are also attempting to refine the 'cost of quality' concepts to incorporate an environmental focus.

* *Product design: develop innovative product solutions* This development will increase the likelihood of staying competitive. Be aware of the numerous environmental criteria affecting product design - especially as they are continuously evolving. The approach to product design should change to include environmental needs - how products fit into the overall lifestyle of society should be considered and revised. As an example, should automotive companies become transportation companies that support/own public transport

and lease cars for weekend travel versus encouraging people to drive to work every day? Responding to this kind of question requires making an assessment of societal needs for products or services, and the cost/benefit ratio to society and the environment. This approach goes beyond the current thinking, which only includes such criteria as materials selection, reusability, ease of disassembly, etc.

* *Communications: provide better communications with environmental drivers* Communicate to the degree that the entire industry can improve decision-making and gain sustainability. Customers, governments, pressure groups, the general public, financiers and functional groups within organisations require good communication tools. Moreover, effective *internal* communication is crucial to achieving the dual goals of better environmental practice and becoming more competitive. The entire organisation needs to understand the company's impact on the environment, the resultant economic effects, and what opportunities are available for meeting both objectives.

* *Organisational structure: become a change oriented, environmentally-proactive organisation* An organisation where the environment is viewed as an investment (opportunity) rather than a cost (compliance issue). There is also the need for a global, as well as local, understanding of environmental pressures and legislation. A change in organisational structure may be necessary in order to achieve a better integration of the company's environmental strategy and its overall business strategy.

* *Training and education: make sure that management, line employees and customers are environmentally aware* Awareness includes knowledge of the problems that certain processes and the resultant products and services create for society. They also need to be educated about relevant legislation and how it will affect business and lifestyles. Final consumers and intermediate customers must be trained about how to use and dispose of products properly, as well as how to improve their consumption patterns.

* *Measuring performance: develop better measurement tools, reporting and auditing systems* Companies require such systems so that they can communicate more productively and be able to set reasonable stretch targets, including environmentally-based decision-making.

Matching Infrastructure with Strategy

In the bottom half of Figure 4.1, a detailed set of the processes, systems and resources to be used for *achieving* the required strategic responses is given.

Competency Organisational competency is the link between the two halves of the diagnostic. An environmental competency is defined as 'having collective learning in a specific area; being able to coordinate skills, technology and processes; and successfully integrate the organisational culture to breed excellence and continued growth in an environmental context.' The ability of an organisation to develop new competencies (and the underlying infrastructure) limits the strategic responses that can be implemented, as developing competencies provides a company *with the means* to respond strategically to its environmental challenges.

Distinctive environmental competency By having a unique set of skills and capabilities, a company can achieve a definite competitive advantage in its industry or market. Novo Nordisk developed a distinctive competency in external communications well before publishing its first environmental report in 1993. Novo responded to an attack on its use of enzymes in the detergent industry by holding a series of meetings with stakeholder groups, both external and internal to the industry. This experience led to gaining leadership status in listening to and learning from the concerns of stakeholders, as well as establishing an organisational structure that responds proactively to environmental issues in all its business interests. Novo's distinctive competency is the knowledge and reputation it has achieved in environmental issues and its impact on the industry - thereby giving the company a head start in new research and product development.

Essential competency Specific know-how is typically needed to be a player in a particular business, i.e., a water supply company must meet the regulations for water quality. *Plain competencies* are routine capabilities such as payroll systems and accounting functions - they are often candidates for outsourcing as they are usually not a major 'value-adding' activity. Companies in the MIBE Project identified the following environmental competencies as being necessary for strategically responding to environmental discontinuities and the changing expectations of customers and other stakeholders:

* Company Q needs to develop the ability to quickly respond to customers' environmental needs in developing products, services and new business.

* Company R needs to develop a competency in total life-cycle product/service design.

* Company S needs to become an expert in developing and implementing environmental assessment tools.

Processes As shown in Figure 4.3, *processes* are the sub-components of a competency. They are essentially the building blocks that work together to form a competency. Processes form a map or flowchart of activities that fulfil certain objectives and can be measured in terms of efficiency against a pre-determined ratio or standard. A process is tangible, whereas a competency is more subjective, including not only processes but the *manner* in which these processes are used and integrated. A *practice* is the way in which the process is managed.

Resources Resources include the people, finances, information systems capabilities, knowledge, time, etc. that are necessary to acquire, make a shift or build a competency successfully.

It is important to understand the specific processes, metrics and resources directly related to achieving a desired competency. Environmental change really means eliminating the gaps between the present infrastructure and whatever is needed to attain the desired competency. *Competency maps* help identify the capabilities, processes and resources that the benchmarking team then must attempt to develop. Competency mapping is a structured format used to help the team define each sub-level of the required competency. The more detailed the breakdown, the easier it will be for the re-engineering and benchmarking team to focus on the most critical processes and systems. The competency map provides a physical overview of the competency-capability-process-resource linkage.

Finally, the team should reassess how well the final *outputs*, i.e., the final processes designated for benchmarking - will affect the overall goods and services provided by the company. Will these changes efficiently address the strategic responses necessary to meet the company's highest priority challenges?

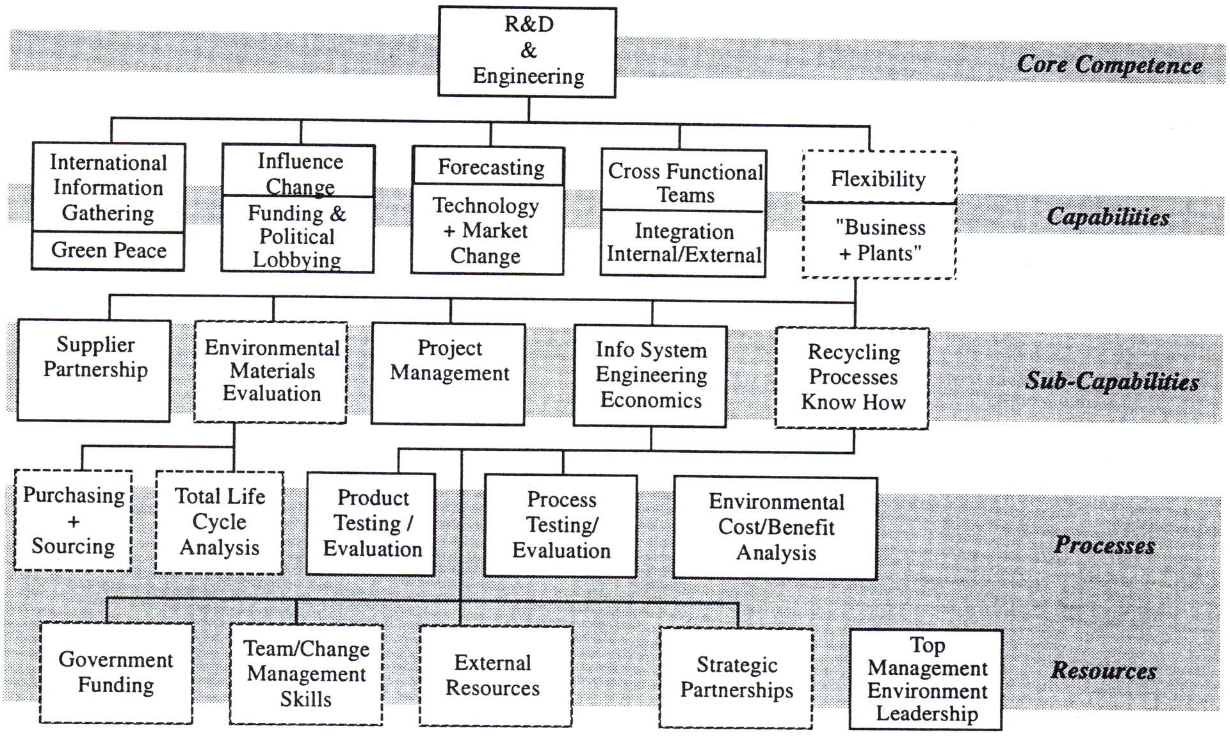

Figure 4.3 Sample competency map: R&D and engineering

Diagnostic Facilitation

Using the diagnostic A cross-functional management team of eight to ten people has proved to be the most effective number for the 'discovery' phase of environmental benchmarking. When the group seeking consensus is larger, the process still functions satisfactorily; it just takes more time. Since environmental issues have an impact on every function in an organisation, it is paramount to receive input from several functional areas. A functionally mixed team is more likely to contribute non-traditional input, which is more apt to lead in the right direction for environmental benchmarking. These individuals will probably not be directly responsible for conducting the final benchmarking activity (phases 2 to 4), but the mix allows the benchmarking process to start with a 'helicopter view'. As well, a good functional mix will ensure less chance of issues polarising early on. In practice, the diagnostic facilitation process has proved to be an 'eye-opening' experience for each of the organisations that have experienced it. The priorities that are generated are not always the ones that most environmental managers expect. One user of the diagnostic remarked:

> The benchmarking diagnostic was extremely helpful in focusing our team on the real issues - the process helped force us to define the competencies necessary for being successful and to think about the different components, in a neutral forum, where we could discuss them and converge on our choices.

The session is usually facilitated by one or two people who guide the group through each component of the diagnostic using the 'post-it' process (described below). This process has been a successful way to generate unique ideas that are free-flowing and non-judgmental. It is important to conduct the session in this manner, so that preconceived values and the priorities of particular individuals do not stifle other people's ideas.

Post-it note process Each individual is given a pad of 7.5 x 12.5 cm (3x5 in) post-its (self-sticking notes); only one thought or statement addressing the component of the diagnostic being analysed is written on each post-it. The post-its are then attached to a blank flip chart at the front of the meeting room. When all ideas and thoughts related to the particular subject have been placed on the chart, the process of confirmation begins.

The facilitator clarifies the statement written on each post-it note with the entire group. After each statement is read aloud, the post-it is transferred to the second flip chart. If someone does not understand a particular statement, it is clarified by the author before it is moved to the second flip chart. When all the statements have been clarified, the members of the group are requested to collectively arrange the statements by categories. (An example is shown in Figure 4.4.)

When all the statements have been arranged into groups of similar concepts or ideas, a title is given to each grouping of post-it notes. In practice, this step is sometimes a bit difficult but is usually creative. Typically, the titles have to be revised a few times. Subsequently, some post-it notes may need to be shifted from one group to another. After titles have been assigned to each grouping, the participants individually rank the relative importance of these groupings. The ranking scheme is structured so that each individual can designate four priorities: '4' refers to the most important grouping and '1' is for the lowest priority item. The individual rankings are then added to form an overall priority evaluation. This step is especially important as it determines the *overall* priority amongst the team members, i.e., it generates a pareto analysis, with two or three categories weighted much higher than the others (Figure 4.4). At this stage, the facilitator - together with the team - devises a statement that captures the overall message or conclusion of the analysis. The wording of this message is adjusted until everyone on the team is satisfied.

The same sequence is followed for each component of the diagnostic, beginning with either the discontinuities or expectations - the choice being up to the team. The discontinuities and expectations are derived independently of each other, with five flip charts ultimately being produced: internal discontinuities; external discontinuities; customers' expectations; expectations of the customers' customers; and other stakeholders' expectations. The next step, using the same approach, is to develop the strategic responses and competency requirements segmentally by linking the ideas on the post-its back to the challenges. The strategic responses generated at this point will address those expectations and discontinuities - based on the first five rounds of post-its proposed which take the highest priority. The competencies to be developed must then correspond to these top priority strategic responses.

Generating two streams of strategic response and competency charts has sometimes been valuable to the team: one set indicates the actions currently underway in the company, and the other citing those that need to be taken or developed. This part of the exercise can help the group realise which strategic responses and competencies fit the existing corporate culture more closely. It also provides a sense of balance and a morale booster for the group; by including positive aspects, they are less apt to spend the entire day focusing on the company's shortcomings.

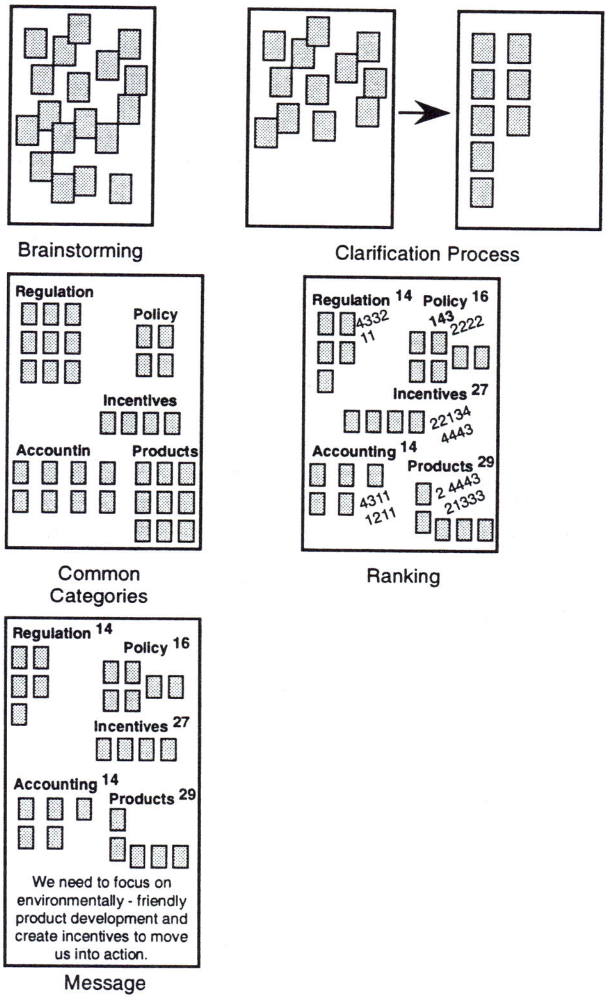

Figure 4.4 Pictorial of the post-it process

Before moving to strategic responses, the strategic intent - the environmental mission statement - of the company is reviewed. A very interesting interaction between the strategic intent and strategic response can often be observed, which results in a much better understanding of the strategic intent or environmental mission. Frequently, the original mission statement is too vague; by interacting between strategic responses and strategic intent, the mission statement becomes sharper, more detailed, more clearly focused - and even perhaps defines what is *not* part of the environmental mission.

A new stage of the facilitation process begins after the competency fulcrum of the diagnostic is reached. The lower portion of the diagnostic is used to build competency maps (as shown previously in Figure 4.3). First, the needed competency is placed at the top of the chart; the team then recommends the first layer of capabilities that the company must develop in order to support the competency, again using post-its.

The next layer of the map cites the processes that are required to develop the capabilities listed in the previous layer. It is usually not necessary to expand every capability. The team can decide to focus on certain more critical capabilities or on those that can be developed more quickly because of their fit with the company's culture and business strategy. It is at this point that the key processes for benchmarking are identified. Throughout this procedure, it is advisable that the team constantly reconfirm their focus on the particular competencies that will lead to improved environmental performance and a higher level of competitiveness.

Completing the upper portion of the diagnostic typically takes six to eight hours, while the bottom half requires an additional four to six hours. Two days should be set aside for using the diagnostic tool - the procedure and the personal interaction are intensive. At the beginning of the second day, the team should review the competencies being proposed for development in order to ensure that they do indeed address the discontinuities and expectations of highest priority. As well, it is good practice to reconfirm, as the competencies are being developed, that they continue to fit the corporate culture and business strategy.

A critical outcome of the competency mapping activity is to determine the exact processes that one wishes to benchmark. That is, a red thread can be drawn from the prioritised challenges to the prioritised strategic responses (consistent with the strategic intent) that then leads to the necessary new competencies to *achieve* those responses. Thereafter, competency mapping develops the detailed infrastructure necessary to attain the competency. The key requirement is to identify those processes and systems necessary to build the competencies. This is where the subsequent benchmarking efforts need to be focused.

The Discovery Phase at Sulzer

Sulzer began its environmental benchmarking journey with the benchmarking team spending one full day at the company's headquarters in Winterthur, Switzerland. This meeting focused on going through the top half of the environmental business transformation diagnostic. This initial team was comprised of eight individuals from three different country operations, including representatives from the strategic development group. The development group continued as supporting members for the study - the only individuals from the initial session who remained a part of the final environmental benchmarking team.

Challenges The Sulzer team identified 12 internal discontinuity challenges that the company was facing, as well as 28 external discontinuity challenges (see Figures 4.5 and 4.6). For the internal discontinuities, the priority assignment by team members focused on the second grouping, entitled 'constraints', with a scoring of 24. The summary message captures the concern that organisational and financial constraints limit the company's ability to meet environmental mandates. For external discontinuities (see Figure 4.6), the major challenges are assigned to the grouping entitled 'rules and regulations'. The shift/increase in regulations from water to air to soil and the growth in environmental liability for companies and managers were regarded as particularly important challenges. As well, this category is viewed as a major challenge for both Sulzer and its customers, as they are increasingly being asked to comply with changing environmental mandates.

Awareness (13)

- Environment is an integral part of TQM
- Visible commitment to environment signed by all senior executives
- Educational campaign to raise commitment of corporate staff to the environment
- Written corporate commitment to lead to sustainable development

Constraints (24)

- Culture/organizational constraints inhibit sales of totally integrated solutions
- We are asked to deliver on the environmental mandate without additional funding to do it (generate the funds yourself but not at the expense of the bottom line)
- Conflict between short-term financial results and long-term environmental solutions
- Sometimes our internal high quality standards in plant designing do not allow Company 5 to compete in local markets

Stimuli (11)

- "Innovation award" for realized environmental projects
- Corporate stimulus for increased debate about the environment within the business, "pro-activity"
- Mid-range planning must include environmental dimensions
- Company views the environment as a good business opportunity

Message: Organizational and Financial Constraints Limit Our Ability to Meet Environmental Mandates

Figure 4.5 Sulzer's internal discontinuities

Rules & Regulations (24)

- Global and regional harmonization of rules
- Non-uniform implementation/application of environmental mandates
- Environmental concern has shifted from water to air to soil (residues)
- "Watchdog" capabilities are improving – Increased growth of environmental intelligence
- Growth in environmental liability of companies and managers
- Approval processes are more difficult (regulatory)
- Shift from "end of pipe" to "in process" treatment

Environmental Reporting and Standards (6)

- Pressure from media (FORTUNE) to be amongst the best "eco" performers
- Growth of "green" funds following corporations with good environmental performance
- Environmental performance is an important criterion for lending institutions and investors
- National environmental management standards starting to "pick up", customers will expect companies to comply with them
- Increasing pressure for more detailed environmental reporting
- Increasing pressure to have an operating strategy document on the environment

Costs (13)

- Trade-off between growth, and environmental costs for countries in development
- Shift from "cost thinking" towards "investment thinking"
- Additional engineering costs for construction approval documentation
- Fear of new technology coming along that will make a current investment obsolete
- The poor are getting poorer and cannot afford environmental protection
- Customers want help with financial issues
- Economic pressures cause "backing off" from environmental standards
- "Green solutions" are perceived to be more expensive

Change in Market Structure (3)

- More information at every stage of projects to be given to customer/authorities
- Globalization and regionalization of customers
- Gaps between what people say and what they do
- In certain sectors, peer pressure is driving environmental initiatives/expectation of performance (e.g. VW and the recyclable car)
- "Responsible care" program of chemical industry (worldwide)
- Environmental pace-setters (e.g. Germany, California)
- Environmental "Perpetuum-Mobile"

Message: Our External Pressures Stem From Existing and Changing Rules and Regulations

Figure 4.6 Sulzer's external discontinuities

Meeting Tomorrow's Needs (11)

- The need to be able to meet stiffer future requirements
- Applied processes should not create new problems in other areas (or future)
- Concerned customers are thinking now about the higher environmental requirements that will exist in the future
- Want modular design that enables upgrades
- Some pioneer customers are willing to have lower pollution limits (investment of more money)
- Industrial customers expect more flexible plants, i.e. wider feed parameters, more versatile
- Consideration of all waste streams and anticipating future pollutants
- Customers want turnkey design, construction, and operation of plant

Cost/Performance Pressures (15)

- They need to meet green legislation at lowest cost
- Lower price levels
- Expect lower emissions at lower costs
- There is no change between environment – or other business; they buy what they can afford
- Plants have to meet or exceed today's regulations at lowest possible price
- Performance of waste treatment according to new emission regulation but:
  - for the lowest price
  - no more than required
- Customers, shareholders expect that "environmental problems" will not slow down their "development" rate

Pressure Groups (7)

- Raise green profile to public and customers
- Help in eliminating opposition – technical and political support
- Help to deal with activists
- Attract "green" investment
- Ability to demonstrate benefits from extra performance

Meeting Today's Needs (15)

- Plants have to have a high degree of safety in normal operation and in emergency situations
- Better references, i.e. proven plant experience
- A lot of customers do not want to be pushed towards lower limits of pollution

Message: Cost-Effective Solutions With Tactical Support That Meet and Exceed Today's Requirements

Figure 4.7 Sulzer's industrial/municipal customer expectations

Sulzer has a wide array of customers, therefore the team decided to focus specifically on the company's industrial and municipal customers. As shown in Figure 4.7, Sulzer identified 23 customer expectations which were grouped into four categories. The summary message for customer expectations focuses the company on cost-effective solutions that not only meet current environmental requirements but exceed them. This is in line with the Sulzer groups image of being a technically excellent organisation. Figure 4.8 captures the 'other stakeholder' expectations, here the diversity of products that Sulzer produces is reflected in the 34 different expectations listed. The team grouped these expectations into 8 categories - for Sulzer, frequently there are political issues affecting its designs, as the systems, machinery, etc. that the company designs/commissions (i.e., incinerators, thermal power systems, nuclear services, valves, etc.) are often partially funded by government agencies, this creates both financial and technical expertise expectations from various stakeholders. The 'greens' - environmentalist groups - and local communities are also identified among its more important stakeholders. The concluding message is consistent with the one seen in Figure 4.7. The primary focus is cost effectiveness - as well as technical expertise and reliable operations - over an extended time frame.

The overall message derived from reviewing these challenges confronting Sulzer is that the company needs to provide technically-excellent value-added solutions that will meet both today's and tomorrow's environmental standards in a cost-effective manner, according to the perceptions of its customers and other stakeholders.

Technical Design Excellence (14)

- Plant operators want protection against legal threats
- Proven solutions for operators (no R&D on site with risks)
- Plant operators wish to have a show-piece "of latest" technology
- Consultants are:
  - looking for innovations
  - politically concerned
- Greens are interested in getting technical support/ designs to meet their demands

Reliable and High Performance Operations (10)

- Health and safety of future generations
- Greens and communities expect higher degree of residue recycling
- Clean and safe community
- Green guarantees for consumers
- Community demands highest level of safety
- Health and safety for employees

Costs vs User Benefits (14)

- Taxpayers want assurance that their money is properly invested
- Plant operators expect minimum operating costs
- Community does not want to pay for it
- Consultants expect to play a major role in "business" definition

Company 5's Cost/Benefit (6)

- Shareholders: extra money, the "green" payback
- Safe investment, less risky projects leading to assured profits for shareholders

Expectations of Greens and Communities (3)

- Measurable improvement in lower emissions
- Creation of processes for full recycling of waste without emissions and residues
- Solutions other than incineration of waste
- Incineration plants with practically "zero" emissions
- "Zero" level emissions
- Everything to be recycled – without waste to be landfilled

Environmental Systems/ Solutions (3)

- Plant operators want protection against "green" threats
- Solutions that don't create more problems for operators, i.e. landfills, special long-term storage
- Politicians expect integrated solutions for waste problems of a region
- Municipalities and politicians expect high credibility of our company (new technology → high risk)

Third Parties in Decision Making (2)

- Consultants expect us to do their work without paying us for this service
- Consultants do not want us to be their only alternative for a given project
- Consultants look for a reliable company, selling reliable technology

Green Aura (2)

- Employees like to work in a "clean" company – green pride
- Politicians supporting us want fundamental improvements, motivated by the green vote
- Green image and reputation for consultants
- Increasing employee awareness on green issues

Message: Solutions That Are Technically Excellent, Cost Effective, and Reliable

Figure 4.8 Sulzer's other stakeholder expectations

Strategic responses Sulzer identified 56 strategic responses, which were grouped into 11 categories that could be used to address its environmental challenges. The highest priority responses focused on four of these response categories: organisation, policy, product development and measuring performance. The summary message in Figure 4.9 provides the proactive link the group found between these responses - to strengthen its organisational structure (including environmentally-related measurement systems) to meet global requirements more effectively, develop clearer policies with respect to market requirements, and develop a new approach to product mix and design to foster environmental business opportunities. A few key responses indicated that Sulzer needed to strengthen its approach to the market with respect to contracting, consulting and operating capabilities: by finding better financing solutions for customers and operators; by reviewing the ethics involved when working in less environmentally-friendly markets; by improving the means for having a global exchange of information; by enhancing the ability to operate on a global basis; and by establishing environmental targets.

Organization (11)
- Global exchange of information within groups regarding different aspects of environmental technologies
- Open a consulting office/practice
- Watchdog on trends in environmental legislation
- The company has to push local activities even though we are a centralized corporation - to be locally worldwide

Incentives (3)
- Financial rewards tied to environmental performance
- Think more toward strategic investments than immediate earnings
- Financial incentives for good environmental performance
- Provide procedures for reviewing environmental benefits of investments

Measuring Performance (6)
- Environmental reporting (corporate and business unit)
- Establish environmental targets (emissions, waste, energy, etc.)
- Create a measurement tool to evaluate our environmental performance
- Create a company tool to measure return on investment for environmentally related projects

Marketing
- Form industrial alliances, i.e. chemical industry
- Partnership relationship with customers, i.e. to look towards a responsible care program

Costs
- Product engineering for entire life cycle of product
- Condition monitoring over time; re: requirements
- Proven technology to be adapted to local needs
- Lean production, lean organization

Policy (11)
- Change traditional company approach to market; separate equipment manufacturing from engineering and contracting
- Establish a clear policy on how to do business in less-environmentally concerned markets
- Contracting and consulting capabilities to be improved (for integrated solutions)
- Take into account cost effectiveness according to the operator's/customer's perspective to propose the best technical solution, including, if necessary, a financing solution
- Develop a clear policy, according to market requirements and profitability, for contracting, building and operating

Product Development (8)
- Ask customers what they want, find an environmental champion
- Modularize products to match changes in legislation
- Design/engineer products to set the regulations/rules
- Focusing on new processes for residuals treatment
- Identify pioneer customers for new/improved products
- Approval procedures as a new business service, i.e. cash cow
- Push R&D activities because the company is losing leadership
- Be consistent, products that help the environment today should not damage it tomorrow

Working with Other Environmental Stakeholders (5)
- Dialogue with communities and "green" groups
- Try to influence:
  - regulations
  - pressure groups

  through lobbying (Brussels) and supporting influence groups
- Better communication with stakeholders (references, performance, cost effectiveness)
- Invite pressure groups into the company; encourage employee membership to better understand pressure group concerns
- Lobby for global standards and energy taxes
- Environmental lobby group for company

Information Intelligence (4)
- Close co-operation with authorities in different countries in order to anticipate new regulations
- Learn from green intelligence
- Updated legislation database system comparing different rules and external discontinuities
- Improve horizontal communication between company business units
- Pooling internal resources from all company players to get
  - more references
  - better information base

  to support customers
- Marketing intelligence system to spot, interpret, and feedback future steps needed for environmentally sound solutions

Public Relations and Communications
- Encourage green investors
- Make company's ability to create technical solutions to environmental problems more public → corporate image on environment
- Have an environmental department/manager
- Be proactive in influencing policy makers to benefit from our technological distinctive competencies and raise standards
- Company to qualify for green funds and FORTUNE best ten performers
- Create a corporate policy to communicate environmental performance to the financial market
- More publicity of our systems and results
- Company should make greater use of political and financial levers

Education
- Encourage people to think in terms of systems and relationships and not categories and processes
- Influence the mind set of our end customer
- Expose young creative engineers to the "business aspects" of environmental issues
- Promotion of cost internalization philosophy... accept environmental cost for what it is
- Customer "education" on environmental programs as part of marketing

Message: ***Strengthen the Organization, Approach to the Market, and Product Development to Foster Environmental Business Opportunities***

Figure 4.9 Sulzer's strategic responses

Core Competencies The next step was for the Sulzer team to identify the new core competencies that would be required to achieve the identified strategic responses. As Figure 4.10 indicates, 25 competencies were grouped into 5 main categories. The two groups with the highest priority were 'marketing' and 'organisation'. 'Information systems', 'championing the environment' and 'measurement' were the other three categories identified as candidates for competency building. No message was developed at this stage, as the team would not begin the competency mapping portion of the diagnostic during this session. Rather, the team decided it was better to revisit the top portion of the diagnostic and reflect on the growing red-thread of organisational and market approach changes - that would enable Sulzer to make the strategic response needed to address the more important environmental challenges facing the company.

Measurement (4)

- Internal environmental performance benchmarks
- Methodology to measure environmental performance
- Reward systems for creative people

Champion the Environment (8)

- A top executive politically active outside the home market
- Get more influence on opinion leaders and decision makers in the environmental market
- Interdisciplinary cross-functional forum on environmental issues

Marketing (15)

- Strong and competent market approach
- Fast development of new products according to future market trends
- Specific market approach for the environment as a business activity (Market Oriented Contracting Division) and for the environment as a general development opportunity (with measurable performance of product activity)
- Compulsory feedback on marketing experience
- Ability to identify what the customer/market really needs and to question/change our present structure to meet those needs

Information Systems (8)

- Gathering perceptions of market needs within 3-5 years (international marketing information system)
- Corporate network to create database on environmental trends and legislation
- Global "Info System" to capture facts and figures on present environment performance as base for targets (and later, reporting)
- Integrated strategic management of technology systems, a strategy of technology management
- Early warning radar

Organization (13)

- New name for Company
- Integration of marketing and engineering capabilities
- Network should extend to pressure groups, i.e. lawyers, politicians, ecologists, biologists, biochemists, public relations, marketing, geologists
- Provide global service but act locally (central coordination, local action)
- Competencies for:
  - Financing big environmental projects (work management facilities)
  - Operating these facilities
- Build up a structure/ competency for customer plant operations
- Systems design philosophy
- R&D department in Brussels
- Engineering to be done at the source of the environmental problem

Figure 4.10 Sulzer's new core competencies

Competency mapping The competency mapping process - the bottom portion of the environmental business transformation diagnostic - required another one-day session. The Sulzer team began by reviewing the results of the first half of the diagnostic process in a more holistic manner. They held a brainstorming session on how Sulzer's businesses interacted with the environment, and developed a clearer view of how and where they were affecting the environment.

This led to a better understanding of how this particular business unit should be organised - what systems were necessary - to satisfy the environmental marketplace (see Figure 4.11 for the result). This review (or a similar process) is fairly common in our use of the environmental business transformation diagnostic. In essence, prioritising the strategic responses often leads to further development of the strategic intent or mission statement by the group. This step enabled Sulzer to understand more precisely what strategy it should pursue and how it could develop ways to evaluate its progress in that direction.

Figure 4.11 also served as a guide for reviewing the competencies identified earlier. During this session, the relative importance of 'measurement' as a competency was reassessed (note its lower ranking in Figure 4.10). Measurement was now seen as a key factor for rationalising the decision to be proactive in setting environmental targets with respect to IWOs (inputs-waste-outputs). Sulzer needed to develop a competency in *measurement*, i.e., to establish procedures and tools that would enable the company to analyse its IWOs. The goal was to ensure that Sulzer's IWOs were at levels that would demonstrate sustainable practice in line with its goal to conserve the environment while simultaneously providing customer-driven solutions.

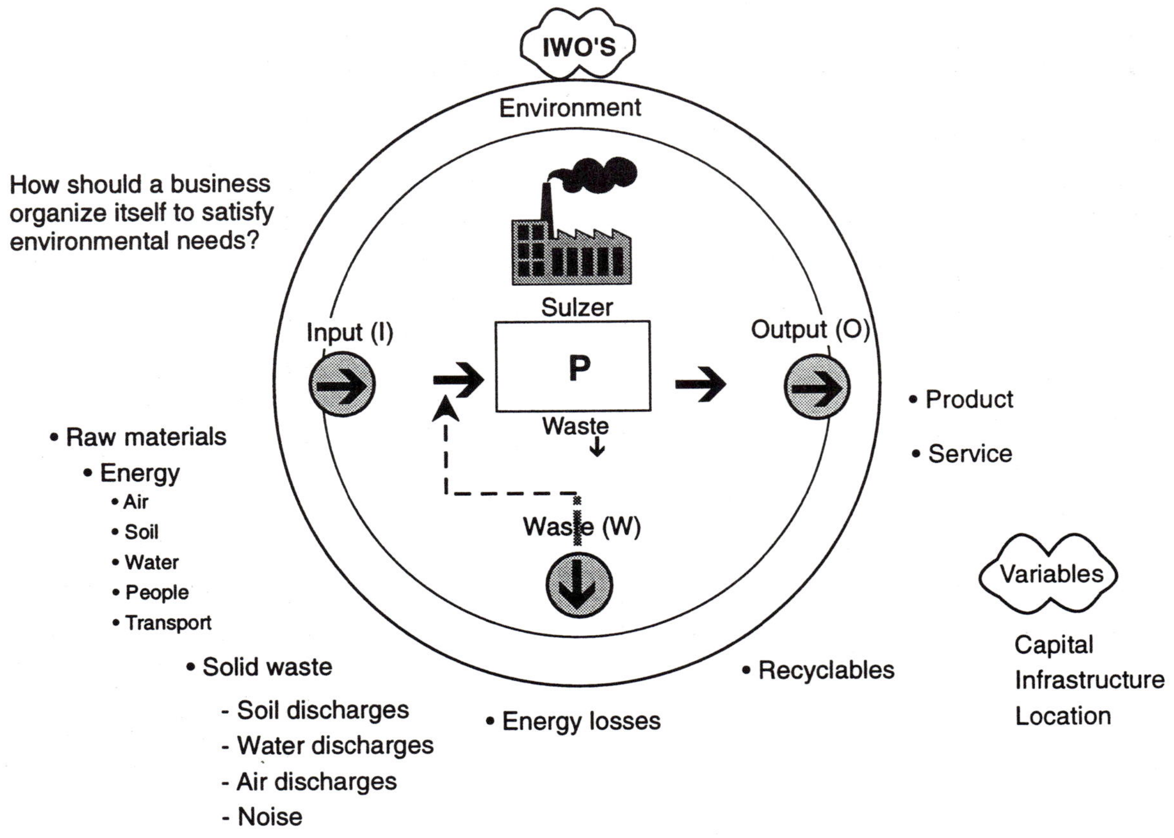

Figure 4.11 Sulzer's operational points of environmental interaction

The Sulzer team felt that this environmental positioning was essential for leading the company toward designing better products and developing better corporate systems that would support continued competitive advantage. The company required a measurement system which could systematically assess the overall environmental impact of its internal operations, and which could be leveraged and applied to the use of its products by its customers. Sulzer's secondary objective - determining what levels of emissions are sustainable - also included assessing the appropriateness of standards and legislated levels of emissions.

The Sulzer team then turned from its environmental positioning to development of a competency map for the newly perceived competency of systematic measurement. The group did not proceed in exactly the same way as shown in our transformation diagnostic, but this is not a problem. The diagnostic process is not intended to be like a straight-jacket - too inhibiting; it is merely a suggested way to proceed, and the *journey* as perceived by the team is an important part of the process. Taking ownership of the process and using the diagnostic procedure only as a guideline can sometimes be more fruitful than sticking to the particular presentation of results outlined previously. The results of the team's effort to develop a competency map are summarised in Figure 4.12. Five supporting capabilities were identified: understand the environment; interpret key environmental factors; evaluate key environmental factors; establish procedures; and finally, measure performance.

The team then extensively deliberated what would be required by Sulzer to *achieve* these five capabilities and the necessary supporting infrastructure (processes and systems). The next steps for this Sulzer business unit will be described at the end of the next chapter (Phase 2: Assessment, How to Select Processes and Environmental Benchmarking Partners?), after a clear description of the environmental benchmarking process.

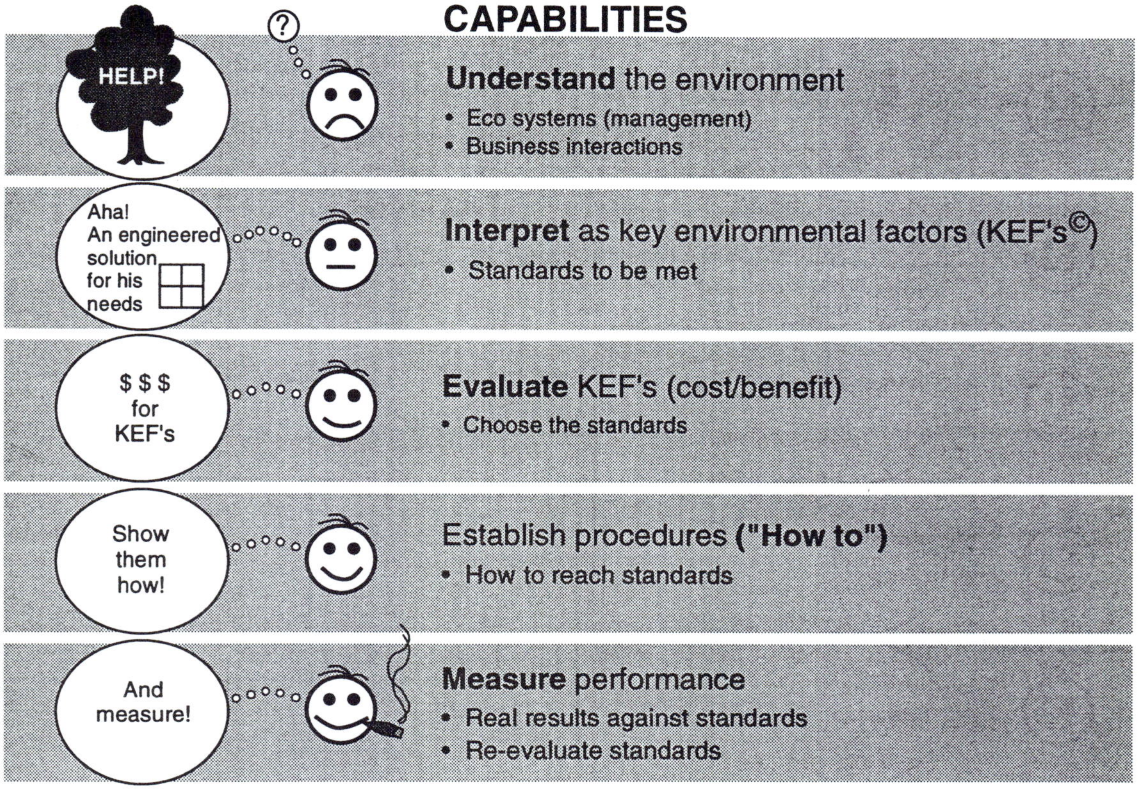

Figure 4.12 Sulzer's capabilities to support a competency in 'measurement'

Chapter Five

PHASE 2: ASSESSMENT
How To Select Processes and Environmental Benchmarking Partners?

This second phase of the benchmarking process focuses on assessment by the benchmarking organisation. Assessment, by definition, indicates a detailed evaluation of key processes and systems, defined through competency mapping. Also critical to achieving the necessary competencies, assessment includes determining the final members of the benchmarking team, selecting the benchmarking sites and preparing material for the visit. This chapter will examine the following six key questions in conducting an effective assessment:

* What is the ideal benchmarking team composition and size?

* Why is it necessary to perform business process re-engineering?

* What are the key aspects of choosing a benchmarking site?

* What homework needs to be done before the visits?

* What are the most common sources of poor performance from benchmarking efforts, and how can they be avoided?

* What was Sulzer's experience?

Selecting the Team

Phase 2 is time intensive; it begins by selecting the members of the benchmarking team. Choosing the team takes place either simultaneously or just after a final decision has been made about which processes to benchmark.

Team members Having the right team members is crucial to a successful study. Choosing the benchmarking team is largely determined by asking who currently owns the parts of this process, and who is the most knowledgeable about it. It is essential that those individuals who will be responsible for implementing and *living the change* are involved in both finding the best new practices and in implementing them within the company. This point can never be over-emphasised, as it is critically important for the process owners or service providers, not just their managers, to contribute to the identification of new methods for doing their work. This approach not only ensures a sense of ownership, but superior and more concise plans and solutions can be developed. In the case of environmental benchmarking, team members could include people knowledgeable about environmental issues and trends in the industry, people with experience with environmental regulations, and those who understand the implications of technological changes on the environment, or people who have an active interest in making the process targeted more 'green'.

Business Process Re-engineering

Business Process Re-engineering (BPR) needs to begin where the environmental business transformation diagnostic ends - with the competency map. The benchmarking team needs to review the competency map in order to clearly understand 'the red thread' - that is, why the particular processes have been designated as critical - and where links to the infrastructure supporting these processes and their practice can be found. The team must have the qualifications needed to study and improve the detailed aspects of the processes - as well as to anticipate future needs. In other words, the team needs to understand the underlying logic for selecting a particular process for benchmarking, why *this* team is the best one to examine it, where *this* process fits into the overall scheme, and its relationship to other parts of the organisation.

Process analysis To accomplish environmental re-engineering, the benchmarking team must systematically 'unravel' the steps of the processes to be benchmarked, as well as the required resources to support them. The term 'unravel' is used to describe the act of physically mapping each operation and all decisions made throughout the process. Typically, a detailed process analysis is drawn - which may be several pages long. It is essential that every member of the team understand how the process works: the actual sequence of events; who is physically involved at each step of the process; resource capacity calculations; identification of bottlenecks or other limitations; decisions taken; time required at every stage; information and communication requirements; the connections between sub-processes and other activities of the business unit; and, finally, the detailed environmental implications at each stage.

An example of this can be illustrated by the experience of the world's largest independent inspection, verification and certification company. This company found that the re-engineering phase of the benchmarking process played an essential role in enabling their team to realise the full implications of a proposed information management system. This company, with a highly decentralised, profit-centre based structure, traditionally communicated along vertical lines. Through the benchmarking process, they identified a critical need for increased communication among country offices in order to become competent in the emerging environmental management systems consulting market. Fundamentally, they needed to develop a new competence, of practice sharing and joint learning in all offices, so as to determine the right experts for a particular contract and the most successful bidding procedures. The team focused its re-engineering on the communications system currently being used - which consisted of the individual exchange of proposals between corporate and the various country managers. The existing culture of vertical communication was clearly an obstacle to improved learning, knowledge transfer, selection of the consulting team for the contracts, and eliminating bias from the contract bidding process. The team developed a computer-based system which would link all affiliates worldwide to support building up competence in timely-informed decision-making at the responsibility level of the field personnel, and which would allow the technical expertise of the company to have a direct influence on the mixture of contracts engaged. The proposed system focused the entire organisation on sharing information and developing an overall competency that could be leveraged in such a decentralised company.

In the 'unravelling process', the logistics and objectives of the proposed computer-based communication system were flowcharted. The need for increased teamwork, team learning and their organisational support were clearly identified as objectives, but were missing in the new system. Achieving this type of support required a change in behaviour as well as acceptance of the new system by the country managers. Moreover, the corporate environmental group needed an efficient and cost-effective mechanism for training and educating all the country managers in better ways to identify new business opportunities. The challenge for the environmental benchmarking team was to find a company that had already implemented a similar radical change in behaviour regarding internal communications and sharing knowledge in a highly decentralised organisation. Then, they would have to make a careful examination of the processes and systems that had supported this competency.

Common environmental processes Through the MIBE Project, we have succeeded in identifying many competencies and their underlying processes that often need to be re-engineered to obtain an environmental competitive advantage. Some of the more interesting ones that have repeatedly recurred include:

* Environmental legislation awareness/influence system
* Product development: life-cycle assessment, designing for the environment
* Material resource assessment/replacement
* Waste minimisation/prevention practices
* Recuperation systems for alternative energy sources
* Environmental accounting systems
* Emergency response systems
* Environmental education and training programs
* Environmental performance measurement systems
* Customer service from an environmental perspective
* Environmental management systems

Deciding Where to Benchmark

The *where* to benchmark question not only refers to a particular company but to what type of benchmarking the company should do to gain maximum leverage. The traditional belief is that the most rewarding benchmarking for bottom-line and customer service improvements comes from either competitive or generic benchmarking (looking at other industries with similar processes).

Internal benchmarking A common practice in benchmarking is to begin by making comparisons among strategic business units (SBUs) in the same company. But, particularly for the environment, this kind of 'benchmarking' needs to be approached cautiously. Finding the environmental breakthrough ideas is less likely when looking internally. However, whether the focus is internal or external, it must concentrate on the practice of specific processes - in order for true learning through benchmarking to be achieved. Internal benchmarking, for all its imperfections, is the place to start and to learn. Every benchmarking team also needs to discover *how* to benchmark - a procedure that is acquired most easily in sister business units. Thereafter, the team knows how to look, listen and learn when conducting their external benchmarking studies.

External benchmarking External benchmarking in the environmental field can be extremely useful, as this is where the 'breakout' ideas are generated. Numerous companies have successfully used external benchmarking to develop innovative systems and approaches for their business systems. But, a much smaller number have applied this approach to integrating environmental aspects into their overall strategy and supporting business systems. Companies need to develop a working understanding of what is meant by the term 'best practice' or 'best-in-class' in environmental terms - in other words, precisely what is excellent environmental performance? As one knowledgeable person in the field professed: 'There is no quick fix to our environmental problems and there is no such thing as a best environmental practice organisation.'

True as this statement may be, it in no way obviates the need to search for best environmental practices - wherever they can be found. There are an increasing number of companies, industry associations and consumer advocacy groups that can be classified as knowledgeable about Excellent Environmental Practice (EEP). This does not mean that these organisations have necessarily reached the ultimate - that they no longer contribute to global warming and ozone depletion. It means that they have been successful at altering some parts of their organisation as they reach toward higher environmental goals - usually with good economic results as well. Environmental benchmarking seeks to find and emulate as many of these EEP practices as possible, focusing on those most applicable to the competency goals of a particular business.

EEP defined Excellent environmental practice is not a well-defined concept - nor is there any one body or even unified point of view as to what constitutes *excellence*. We define EEP as follows:

* *Excellent environmental performance* is a relative value judgement used by various stakeholders (and criteria) to recognise the superior manner by which organisations conduct their business operations and environmental management practices.

This definition implies that each business needs to examine the environmental performance of its own and other business enterprises in relation to the particular stakeholders that influence the EEP assessment - as well as the criteria by which this assessment is made. The involvement of several groups in deciding what 'EEP' is for an industry creates ambiguity. It makes it hard to know with any certainty whether particular practices are seen as excellent by a wide group; it also complicates the linkage between best environmental practices and economic criteria. But, at the same time, this ambiguity permits more creativity in developing particular environmental practices and finding ways to leverage them for competitive advantage. Best practice in the more traditional uses of benchmarking is defined by Robert Camp as 'those practices that please the customer most'.[15] This reflects the strong customer orientation of traditional benchmarking. As discussed further in Chapters 8, it is evident that environmental practices involve numerous stakeholders in the evaluation of excellence - and often customers do not know how to identify a 'good for the environment' product. Thus, the environment becomes an even more challenging field for developing and sustaining distinctive competency.

---

[15] Camp, Robert C., *Benchmarking: The Search for Industry Best Practices the Lead to Superior Performance*, Milwaukee, WI: Quality Press, 1989.

Figure 5.1 illustrates the problem/opportunity. Excellent environmental practice is defined by numerous corporate stakeholders. In Figure 5.1, we present nine separate groups, each with its particular perspective. The key conclusion, in deciding where to benchmark, is that each stakeholder's point of view needs to be understood - as well as the corresponding examples of excellent environmental practice. The fundamental task is to sift through this information, using the competency development objectives as the filter. This *filtering* for choosing where to benchmark can be done by reviewing the prioritised list of environmental challenges, the proposed strategic responses and the need for new environmental competencies - as well as the mapping details.

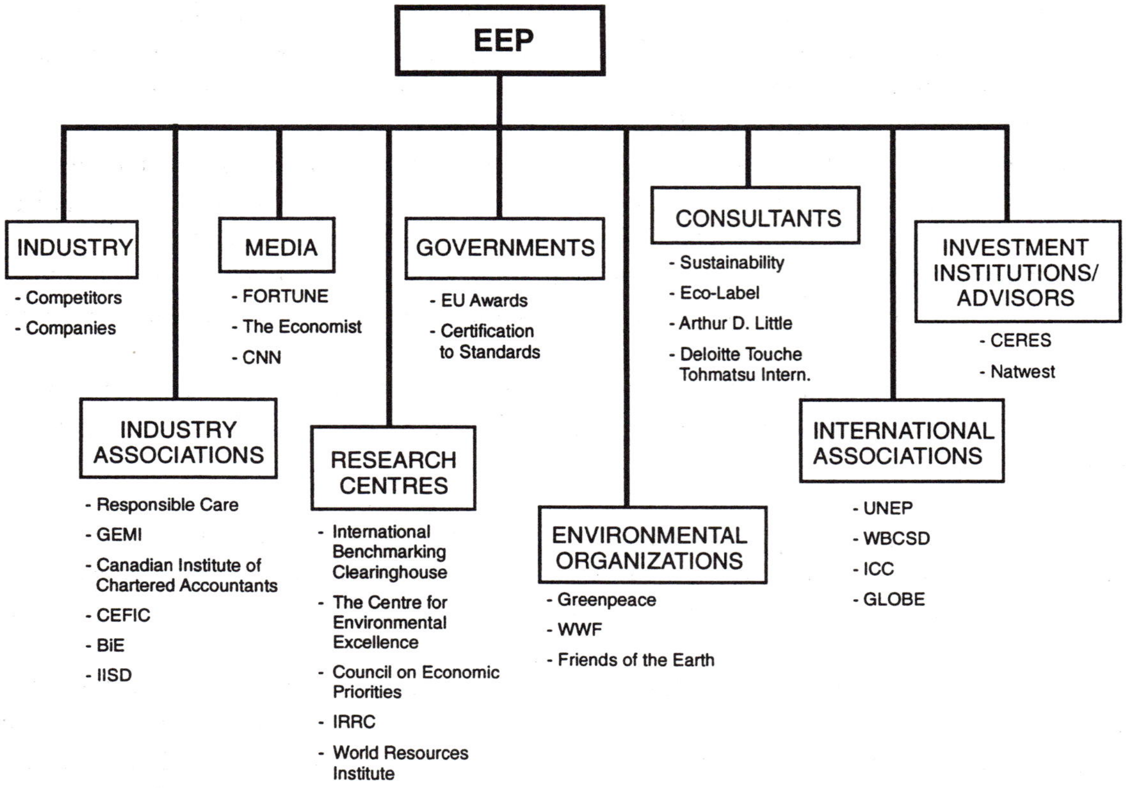

Figure 5.1 Several stakeholders that evaluate excellent environmental practice

Our philosophy is that an organisation can be considered to have excellent performance in one area but not necessarily in another. Moreover, environmental impact has many sociological and ethical undertones. For this reason, it is difficult to assign the status of 'excellent environmental practice' to any corporation on an overall basis. Furthermore, our experience indicates that most firms would shy away from receiving such an accolade, for fear of being singled out for intense scrutiny. The multiple definitions of EEP - as well as the continual need to achieve both economic and environmental objectives - further complicate the issue. For some firms, the only sure way to achieve true excellence would be to spend disproportionate amounts of money and management time on environmental issues. This in turn would mean that the firms would become non-competitive or need to stop operations. This way of thinking is wrong. We have repeatedly asserted that good environmental management also means good business performance. Thus, the challenge that the environmental dimension is putting forward to the business sector is to achieve, *simultaneously*, superior economic and environmental performance results. This presents a challenge for both business and social stakeholders as they seek to develop (and evolve) the guidelines for 'sustainable' excellence.

EEP as a goal At this point, it might seem that achieving the status of 'excellent environmental practice' is like the search for the holy grail - not a reasonable activity for overworked management teams. However, this is *not* the case. Why is achieving EEP so important? This situation clearly parallels the quality movement that began in 1931 when Walter A. Shewhart of Bell Laboratories published 'Economic Control of Quality of Manufactured Products'. It was not until the 1950s that one of Shewhart's students, Dr. Edward Deming, found a way to make Shewhart's ideas *more* than a set of tools. Deming developed a clear business mandate for Shewhart's ideals. Deming's 14-point system for quality professed that quality more closely resembled a philosophy than a technique - but an *operational* philosophy that, if followed, would lead to competitive dominance. Deming's ideals were not readily accepted in the US and thus he looked to the Japanese. The Japanese audience adopted the philosophy with strong conviction, and within a decade the Japanese were no longer 'the great copiers of Western technology'. They became recognised for building superior products that were durable, functional and reliable.

As history is known to repeat itself, a similar *philosophy* can be developed to encompass the economic payoffs from environmental excellence. As shown in Chapter 3, Deming's 14 principles of quality are easily transferred to environmental issues. Leading-edge companies using good environmental practices understand the fundamental benefits from devising an integrated environmental philosophy for conducting their business. The concept is to develop goods and services that will sustain economic growth while meeting society's needs *and* to improve the ecological balance required for the earth to support life.

Organisational fit In the final selection process for a good benchmarking partner, organisational 'fit' may also be a consideration. Implementing the findings from the environmental benchmarking study will affect various areas of the organisation as they incorporate the environment into the business strategy and their daily responsibility - a culture change due to environmental issues often means increased cross-functional teamwork and more open lines of communication. Choosing a benchmarking partner that has a culture close to the ideal state can be critical for developing a successful implementation plan. Similarly, the management structure of the benchmarking organisation should also be considered. If a company is decentralised but the environmental strategy is centralised, choosing a partner with a similar structure might be useful. One of the MIBE companies was in this very situation. The management wanted to benchmark their approach to the environment with a decentralised company that had implemented a centralised system. Matching the approach was more important than whether the benchmarking partner is in a similar industry or business sector.

Doing Homework is Necessary

Before visiting a benchmarking site, a certain amount of homework must be done. The team should know as much as possible before arriving; relevant questions should be communicated to the site beforehand, so that the team can interact with the company confidently and intelligently. Moreover, the team needs to study published information about the company before the visit. There is a great deal of information about companies in the public domain. In fact, several companies visiting IMD have been amazed at the amount of information about their firms found in public databases. As well, with the advent of the electronic information age, the surplus of information available through the Internet computer network can almost be seen as an information overload.

Literature searches To begin the homework step, it is suggested that a preliminary literature search of recently published articles on the specific processes to be benchmarked be conducted. This is a good starting point and method for discovering which companies are considered to have excellent practices. As well, it also permits gathering conflicting views about which practices are better for the environment than others, which ones are more suitable for decentralised or centralised organisational structures, which companies are promoting TQEM, and which ones are integrating the environment in their health and safety functions.

What is the best way to go about conducting literature searches and gathering information? A thorough description of various information sources is presented in Camp's book, 'Benchmarking: The Search For Industry Best Practices That Lead To Superior Performance'. This list has been reproduced, in part, in Figure 5.2 and modified to include environmental sources of information. It is not intended to be exhaustive, but it does provide a good beginning for companies and helps to ensure that they include a wide selection of information.

It is logical to begin the 'homework' stage of the process with the information that is the easiest to obtain. Fortunately, public reporting on environmental programs, initiatives and performance is becoming an annual event for many of the large multinationals and medium-sized companies. As well, many companies consider the environment as an area where businesses should share progress in order to advance the entire industry. Nortel's experience is a good example. Nortel was the first telecommunications company to find an environmentally-friendly circuit board cleaning agent to replace the use of CFCs. Nortel decided to share this technology with other industry players rather than use it as a competitive advantage, because it was known that CFC was a major contributor to destruction of the ozone layer. By sharing this technology, the entire industry was more likely to reach its goal to stop the use of CFCs, thereby meeting the stipulations of the Montreal Protocol. As well, after these players acquired the technology, through the sharing experience, other industries could begin to work on finding substitutions for substances known to cause ecological harm.

Company environmental reports and library CD-ROM searches are two good sources for collecting information. A CD-ROM search by topic will identify the companies being discussed in the latest journals with regard to developing a specified environmental competency. Environmental reports issued by noted companies, as well as other relevant publicly available information, can be obtained - usually by a phone call to the public relations office. These reports provide general information about a company's environmental performance. Typically, the corporate environmental policy, product development guidelines, employee projects and the environmental management system (EMS) are highlighted, which provides clues as to whether a specific company would be a good benchmarking fit. In addition, it is useful to consult some of the numerous reports describing the essentials.

SOURCE	**EXAMPLE**
Internal	
Library CD-ROMs	ABI/Inform
Internal reviews	Environmental audit reports
Internal publications	Company environmental reports
Marketing	Attitudes of the customer base
Computer Networks	Internet
External	
Professional associations	World Business Council for Sustainable Development
Industry publications	Chemical Week
Special industry reports	Economist Intelligence Unit Research Reports
Functional trade publications	Waste and Environment Today
General management	Corporate Environmental Strategy
Functional journals	Environmental Excellence
Conferences/Seminars	The Greening of Industry Network Events
Industry experts/Associations	Responsible Care
Environmental data firms	Investor Responsibility Research Centre, Coalition for Environmentally Responsible Economies
Environmental Experts	
University sources	Environmental Engineering Schools
Library publications	IMD - Guide to Sources of Information on Environmental Management
Company watchers	Ethicscan Canada
Newsletters	Environment Watch
Award systems	EU Award Systems
Environmental think tanks	World Resources Institute
Environmental pressure groups	Greenpeace
International associations	UNEP, WBCSD
Investment Institutions	National Westminster Bank
Original Research	
Customer feedback	Focus groups, surveys
Inquiry service	GreenNet
Networks	Personal contacts
Consulting firms	Arthur D. Little

Figure 5.2 Environmental sources of information

Preparing for the Visits

In all benchmarking studies the most important exchange of information takes place at the 'best practice' visit. The following key methodological issues need to be understood in order to make this process as productive as possible.

Sending prior information The team making the visit should have contacted the other firm, starting at a high level. The contact should eventually lead to a dialogue between the team (with its group of process owners) and the corresponding process owners in the other company. It is important to establish as much interaction before the visit as possible. This step will be facilitated by being totally explicit in designating the processes that are of interest. In fact, it is desirable to send the process analyses ahead of time - that is, the original 'as is' situation (or plans) plus whatever re-engineered solutions the team has generated. This information will give the others a good sense of the journey which has already been travelled, and clarify the issues and the team's hopes for the visit. This information also needs to include a clear identification of the team's *own* feelings about EEP: What is seen as truly important? What are the evaluative criteria?

As well, the benchmarking team needs to develop a questionnaire that will be used as the framework for the visit and the starting point for observing the EEP practices. The questionnaire should thoroughly explore the process that is being benchmarked. Another recommendation is to have each team member develop one question for each component of the process being benchmarked. In this way, the team will have a multitude of questions which can then be streamlined into a very concise set to examine the process from several angles. Inevitably, the more thorough the questionnaire, the higher the quality of the visit is likely to be. A good questionnaire also indicates that the team has a better understanding of the process - as well as the systems and capabilities of the company - being benchmarked.

Prior to making the visit, it is valuable to send the questionnaire to the benchmarking site. As well, biographical information about each team member, as well as present job descriptions, interests and future objectives should be included. These efforts help to ensure that appointments for the visit will be made with appropriate people. It is also useful to provide public information about the visiting company - its products, markets, locations, a brief history, annual reports, environmental mission statement and records of past environmental projects.

Guiding principles In Chapter 6, we will present a more detailed description of how benchmarking is actually accomplished. The rest of this chapter will focus on some guiding principles that we have observed in companies which reputedly have attained EEP in some of their business practices. Our research in the MIBE Project has found several characteristics that EEP companies share to some degree. Many of the companies have adopted a set of principles with three common goals.

* *Pursuing the concept of sustainable development as an overriding objective to guide all environmental programs.* This lofty objective is more practical than it might appear. As a policy statement that everyone is asked to use as a basis for concrete actions, companies are achieving a genuine integration of the environment into business decision-making at all levels of the organisation.

* *Fostering the belief that the environment is a true source of differentiation for the company.* In response to market forces and consumer demand, many opportunities for differentiation exist; EEP companies have found that satisfying and anticipating customers' environmental needs, understanding the company's environmental impact, developing systems that promote ecological compatibility, and even *setting* the environmental agenda for the industry are not just good citizenship - they result in good business.

* *Believing in continuous improvement.* Adapting the lessons learned from the quality movement and applying the maxim 'good can always be better' to environmental performance promotes incremental improvements. More importantly, it establishes a way of thinking where everyone takes some part in the environmental learning process. The attitude is not to sit back and wait for the 'big bang' technological leaps or for new legislation and regulation before attempting to meet environmental challenges.

These three principles are important, but they are not sufficient by themselves. We have isolated four other attributes of excellent environmental practice. Although all of them apply to environmental benchmarking, they also have more general applicability - to the way in which best environmental practices are integrated on a day-to-day basis.

* *Attitude*, the first of these attributes, clearly must be transmitted throughout the organisation, but it has to emanate directly from a commitment to the environment made at the top level. Attitude is permeated or created when: the environment becomes an important part of most public communications; internal company awareness and recognition programs are in place; there is wide participation in voluntary environmental improvement projects; employees are encouraged to take an active role in promoting such projects -in the company as well as in the community. As well, these short and long-term strategic goals, i.e., enhancing environmental performance and being responsible for meeting environmental goals and objectives - are practised throughout the organisation by all functions.

At the 3M company, commitment to improve its environmental performance and be a leader in environmental solutions began in the mid-1970s. In 1975, 3M's Board of Directors signed a corporate environmental policy which focused on the following objectives: solving their own environmental pollution problems; preventing pollution at the source; developing products with minimal impact; practising reclamation; meeting all regulations; and pledging to work with governments and other organisations engaged in environmental activities. At 3M, the CEO sets the environmental performance targets and the rest of the organisation works in teams to meet these targets. The commitment made by 3M enables all its employees to practice each of the 'attitude' attributes described above.

* *Awareness*, the second attribute, is directly related to the method by which a company communicates and builds relationships both internally and externally. Internal awareness is often raised by making a serious commitment to education and training programs. The best ones not only provide technical know-how, but also include public relations and active decision-making where tradeoffs are evaluated and defended. Best practice also means having a formal environmental communications structure that includes monthly or quarterly publications, newsletters, ecological games, etc. Leading firms have also established internal recognition systems to increase awareness and action.

Raising external awareness goes significantly further. Opening a dialogue with outside stakeholder groups can lead to a better understanding of immediate and long-range concerns, as well as to a thorough organisational understanding of how a company is viewed by outside groups. However, these groups often have valuable ideas; and, even when their ideas are *not* good ones, they exist nonetheless and need to be addressed. Methods for raising external awareness include environmental reporting, business chain management, stakeholder seminars and actively encouraging employees to participate in these outside stakeholder groups.

Dow Europe made the conscious decision to raise its *internal* awareness before dealing with external awareness. As a result of this effort, the company now considers its employees to be the driving force behind its Responsible Care program. Dow Europe devised an Environmental Care Award program where the company awards five prizes of $2,000 each year for the best environmentally related project Europe-wide. In 1993 Dow received over 100 applications for these five prizes.

Dow Canada has been noted for its efforts in raising *external* awareness. The company is one of the few organisations that is ranked as being in stage '4' of UNEP's environmental reporting assessment scheme.[16] As well, it is responsible for developing the chemical industries voluntary Responsible Care program. The company also initiated the development of a voluntary scheme for Canada, ARET, similar to the 33/50 program[17] in the US and invited environmentalists to participate in its formulation.

* *Accountability* is the third attribute seen in companies that have been particularly successful in integrating the environment into their routine ways of doing business. Fundamentally, accountability involves a shift in the way a company assesses *its* responsibility for having an impact on the environment. An essential underpinning is having an environmental accounting system - one that is encompassing, open to all and continually improving - both in terms of performance and the scope of the system. Accountability also includes using environmental indices in public reporting, and clearly reporting targets and objectives.

An excellent example of accountability is seen at BSO/Origin. The president of this Dutch computer company developed an environmental accounting methodology which estimates the monetary cost of the following: air emissions, waste water, waste costs and the extracted values of the company's impact; the theoretical costs of having to repair the damage from that impact; and the cost of alternatives and the value of its environmental expenditures. After generating the 'Net Value Extracted' from the environment, the company subtracts that amount from its overall 'value-added' total.

* *Action* is the fourth and last attribute related to achieving excellent environmental practice. The term 'action' refers to how companies prioritise their employees' efforts. Is the environment the first priority or a secondary one? For example, in product research and development, are environmental criteria seen as opportunities or as constraints? Action involves a proactive stance toward developing industry partnerships and/or participating in

---

16 UNEPs Technical Report No. 24: Company Environmental Reporting as described in Chapter 3.

17 The 33/50 program is a voluntary program launched by the US Environmental Protection Agency, and was initiated in favour of self-regulation. It called for companies to reduce their 1988 emission levels of 17 high-priority toxic pollutants by 33% by 1992, and by 50% by 1995.

industry-wide environmental associations; it also means understanding and taking part in the legislative process for the environment in a non-adversarial role as much as possible.

These four attributes for achieving day-to-day excellent environmental practice might not be essential in the selection of benchmarking partners when the goal is narrow - that is, for gaining an understanding of how particular processes are practised. But, at some point, the goals become broader: companies wish to learn from an overall perspective how to develop environmental practices as a competitive weapon.

Poor Performance in Benchmarking

By this point, we hope we have provided convincing evidence that environmental benchmarking is an excellent way to improve both a company's environmental performance *and* its competitive position. But, benchmarking is an art that needs to be practised intelligently in order to achieve best results. There are several common mistakes that should be avoided - to keep the team from only accomplishing 'recreational benchmarking'.

'Feel good' trips Feel good trips occur when firms undertake benchmarking under the presumption that they are already doing a fine job. Then, the real purpose of the benchmarking trip is merely to convince themselves (and others) that this is indeed the case. The team making this mistake tends to focus primarily on finding things that the company being visited does not do as well as themselves, in addition to reasons why they should not or could not adopt any concepts that are in fact superior. Typical responses are: 'Well, they certainly do not have as good a process for monitoring water pollution as we do;' or 'The EPA would never let us get away with that simplified reporting system.'

This kind of bias needs to be systematically examined and rooted out. Benchmarking teams need to adopt the attitude that *every* visit can yield at least one good idea and, if they are not finding ideas that could lead to internal improvements, then they are not functioning as a benchmarking team as well as they should. Particularly when it comes to environmental benchmarking, there will be great differences in problem definition, the approaches used to improve performance, and the measures of effectiveness. The benchmarking team needs to thoroughly understand the perception of the problem in the company being visited. This perception should itself be systematically compared with what exists in their own company in order to understand the differences and to determine what lessons can be drawn.

Environmental benchmarking needs to explore the 'why', 'what', and 'how'. The 'why' is primarily a question of measures - that is, why is a specific step used in the process - why are certain variables being measured to assess a particular environmental issue - and why are these measures appropriate? The question of 'what' is always the most interesting. It relates to the detailed practices of processes which, in fact, allow the company to achieve its measured results. But, the 'how' question is also of significant importance to environmental benchmarking: How does the culture in this company support environmental improvement? How can aspects of this culture be copied or adapted? All three questions need to be part of the benchmarking assessment; good answers ensure that the benchmarking efforts will produce more than just 'feel good' results.

Vague questions/goals Another common problem in environmental benchmarking is having only vague questions or goals when approaching another firm. For example, the stated purpose of an environmental benchmarking visit usually should not be something as general as 'we wish to see how you respond to environmental issues'. At the same time, the team should be willing to take advantage of unexpected opportunities when visiting another company; it is certainly possible to see something very interesting that was not anticipated. But, being quite clear on the objectives of the benchmarking visit will make the company aware that the intent is serious and will facilitate putting together a team that will be more likely to gain something from the visit.

Too broad a scope A related problem is that the scope of the environmental benchmarking visit is sometimes defined in terms that are too broad. Asking a chemical company to benchmark the way it achieves good water pollution control is much too broad. Even asking for best practices in waste water treatment might be too broad, since the particular approach to waste water will vary depending on the chemical pollutant involved.

When the benchmarking partner is approached with too broad a scope, it is because no one did the internal homework required. The need to clearly understand the visiting company's own processes - and in fact to have re-engineered them - is basic to benchmarking. It is only at this point that one is clear on what they want to know, i.e., what new knowledge needs to be gathered from benchmarking.

Focus on metrics not processes The 'why' issue in benchmarking is always one of the first issues to come up. Why do firms perform according to some measure - such as parts per million of some pollutant? Although this interest certainly has merit, the much more important question relates to the detailed processes and systems that allow the company to achieve real measures of effectiveness.

This problem is even more acute when databases are accessed and some organisational unit is told that they are X percent worse than the best practice unit cited in the database. This kind of information is not very helpful, since it does not tell the recipients how that better performance is in fact being achieved. All too often, the result is denied by the recipients, or they believe that the measurements must have been calculated differently or there was some special application that does not relate to real-world practice. The *only* way to truly overcome this problem is to focus on the explicit practices of processes - and to have the existing process owners do the analysis. They are the ones who need to be convinced that the observed practices can indeed be applied in their company.

Lack of team commitment It is the benchmarking team who is most ideally suited to understanding the potential for improvement and the agenda required to implement the resultant changes. As stated several times, the team needs to include process owners - those who work with the existing processes on a daily basis and have the most intimate knowledge of how they work. These people are in the best position to understand the change requirements, to convince others of the need for the changes, and to lead the implementation efforts.

But, sometimes using this team approach just does not work. The team is not properly motivated to take the steps that would make implementation happen, they lack the power or management commitment, or they are just too busy with other things - and are not given enough time to implement the new ideas. In order to adopt many environmental improvement practices, it is necessary to make changes that cross significant organisational boundaries; frequently, the necessary commitment has to be at a reasonably high level in order to ensure that implementation really takes place. Failing to implement what was learned through an environmental benchmarking study is truly unfortunate, because not only are the benchmarking efforts wasted - the future use of benchmarking in the company is also undermined.

Homework not done Failure to perform a reasonable amount of homework before visiting a company means that valuable time is spent during the visit gathering information that could have been obtained in other ways. This information needs to be shared amongst the members of the benchmarking team that will make the visit, then discussed and compared with the benchmarking objectives so that clarification questions can be thought out beforehand.

Wrong benchmark chosen Another mistake that is sometimes made is to select the wrong benchmark. As noted in the environmental business transformation diagnostic, we believe that it is critical to follow the dictates of integration and consistency, so that the processes and systems selected for re-engineering and benchmarking are the most important ones for achieving the firm's strategic intent with regard to the environment. Another firm may be proud of its accomplishments in a particular environmental variable, but the key question is whether that variable is important for achieving the company's environmental objectives.

The benchmarking team needs to understand where a major improvement in a particular environmental variable will allow the company to establish a competitive edge. And these variables are continually changing. Consumer interest in packaging waste or using recycled material may be a critical issue one year and then not be what differentiates competitors at another time.

Failure to look at other industries Too many firms believe that the benchmark for environmental performance is set by the firm with the best performance in its industry. But the true outpacing achievements are found not by meeting the current best performance - but by establishing a new norm that is significantly superior to this level. These are found by looking at environmental practices in other industry sectors.

Many firms interested in finding best practices look carefully at those industries where environmental issues have a long history. These industries have had numerous experiences with improving environmental performance - including how to mobilise all employees, how to respond to environmental pressure groups, handling the issues associated with developing worldwide standards, pursuing cooperative efforts to define environmental practices, and sharing responses to significant environmental accidents - such industries serve as a valuable example.

Inadequate follow-up activities Another unfortunate mistake made by too many firms is not following up the benchmarking efforts with the right action programs. The fundamental issue here is that firms often *intend* to embark on major environmental improvement efforts, but after the studies are completed and the decisions regarding the results are formulated, the firms then may decide that they just do not want the resultant exposure.

Every benchmarking study needs to proceed to a logical conclusion. If not, those who developed the study become greatly frustrated, and the company's environmental programs can be significantly held back. Sometimes, lack of activity is caused not so much by caution on the part of the company but because the project itself is just not carried through to conclusion. Environmental projects, by definition, usually have an impact on a fairly wide range of individuals, or process owners. Coordinating the actions of disparate groups is not easy. But, the costs of not following through are very real for those who have put their efforts into the benchmarking process and then do not see any tangible results.

Sulzer's Assessment Journey

In Chapter 4 we saw that Sulzer stopped its competency mapping process after identifying the capabilities required to build up its competency in 'measurement'. Sulzer's management then decided that they could not progress further in the process without first involving a manufacturing site. This decision was made in part because one of Sulzer's strategic development group's mandates is to design concepts that are practical for line management. Sulzer tries to avoid the 'corporate cookbook' or 'handbook' approach. To support this philosophy, the team modified its approach to reflect more practical questions such as:

> If I am a plant manager of Sulzer X, and I want my plant to be more environmentally-friendly, what should I do?
>
> We really put ourselves in the shoes of the manager and asked ourselves, what can I do to improve my environmental performance? We began to look at the situation from the eyes of the plant manager.

At this point, the corporate delegate for quality and the environment, who was a member of the initial environmental benchmarking team, suggested adding a Sulzer facility manager to the final benchmarking team. The fundamental idea here was that if Sulzer was to be able to market environmental solutions proactively, they needed to develop first-hand experience in how a line manager would evaluate environmental progress. This required having a Sulzer facility that would serve as a pilot. The facility needed to be small enough to manage and close to Sulzer headquarters. Sulzer Burckhardt, a division of the Sulzer Technology Corporation, was the logical choice as it had an environmental program up and running, as well as being certified at the ISO 9000 quality level. These two efforts could be leveraged in order to gain a better understanding of the environmental measurement process at Sulzer.

> We presented the idea to the division manager, highlighting the measurement competency we were hoping to develop and the goals for the environmental benchmarking project from corporate's perspective. A direct benefit of being involved in the study was that MSB would have another tool to use in its decision-making process.

Initially, the MSB division manager was hesitant about participating, as many programs were already running and all resources were committed. But, he left the final decision in the hands of the plant manager, since it was his time that would be needed and his budget that would cover any costs incurred. There was some concern that the plant manager might not think that his manufacturing processes could cause any environmental problems - since Sulzer was not seen as a dirty business. As well, plant managers tend to shy away from re-inventing the wheel (i.e., are cleaner processes really needed?). This kind of thinking brought the team back to their discussion on legislation. They felt confident that they did not have a problem, but they agreed that it was worth the effort to confirm this impression.

> We got lucky - it turned out that the MSB plant manager was concerned about the impact of acetone on the environment. Even though acetone is not a legislated substance, and MSB does not use significant amounts of it, the plant manager was troubled about using it and was considering replacing it with another chemical. However, he was unsure whether a change would be beneficial. Thus, the goal became clear - to understand (and measure) how acetone affects the environment, and to determine the sustainable levels that MSB can emit.

With a line manager from the MSB plant on board, Sulzer developed a matrix which expanded and connected the capabilities identified earlier as presented in Figure 4.12. Five key capabilities are shown in Figure 5.3: understanding the environment, industry/community infrastructure analysis, measurement and tracking, improve performance, and 'green' accounting. The matrix categorises the capabilities into three groupings: those that are external and that Sulzer does not currently have, those that are internal and exist at Sulzer, and those that are related to 'costing the environment' and are considered new for the industry. The intent of developing this matrix was to review where the benchmarking study could add new knowledge. Sulzer was already using and implementing new measurement and tracking systems internally, as well as internal environmental performance improvement systems. The team decided that the area where they could add the most value was in the external (non-existing) category. Of the two capabilities shown, the decision was made to focus on 'understanding the environment'. How should the appropriate levels of waste from Sulzer's businesses, specifically the MSB's emission of acetone, be set (and measured) to meet environmentally sustainable levels?

	External (Non-Existing)		**Internal (Existing)**		**$ (New)**
Capabilities	**Understand Environment**	**Industry/ Community Infrastructure Analysis**	**Measurement & Tracking**	**Improve Performance**	**"Green" Accounting**
Required Result	Eco-system Tolerance level (Sustainable limits)	Sulzer B.U. Tolerance Level	Emissions per units of product	"How to" processes employee motivation	Costs & benefits to Sulzer and environment
What to benchmark (process)	Defining ecosystems Info gathering Info interpreting Analy. & Modelling	Urban planning Population Densities Business/Commy. Systems	C+I Data logging Sensors Auditing	Education & training Suggestion schemes Maintenance L.C.A. Recycling Substitution	Costing: Earth Services Employees Transport
With whom	E.U. DG 17! ETH Greenpeace WWF Northumbrian Water	ETH Urban Studies Dept Local Council	Innotec Dow Surrey Norsk Hydro	DOW	U.C.L. York

Figure 5.3 Sulzer's capabilities analysis matrix

In this case, the Sulzer team was benchmarking an underlying interpretation process for how one was to 'understand the environment' properly rather than a particular process with environmental attributes. For this reason, the acetone issue, although providing a context for the benchmarking, was not in itself the primary objective. Rather, the benchmarking study wanted to find and compare the ways in which other companies went about understanding how some potential pollutant - like acetone - should be assessed and measured.

In other words, instead of re-engineering, Sulzer needed to *engineer* a process for understanding the environment. Their decision was to consult Entec[18], one of the MIBE companies that had expertise in environmental systems design for assistance in the initial *engineering* of this process, prior to benchmarking with other companies. Sulzer had created a preliminary process for understanding the environment before this meeting to use as a basic structure for discussion and to identify where they lacked information. Immediately following the meeting with Entec, the team drew up the minutes and any relevant charts in order to ensure that nothing was forgotten or misinterpreted. The final result (see Figure 5.4) was a process flow for understanding the environment and what level of impact the environment could sustain.

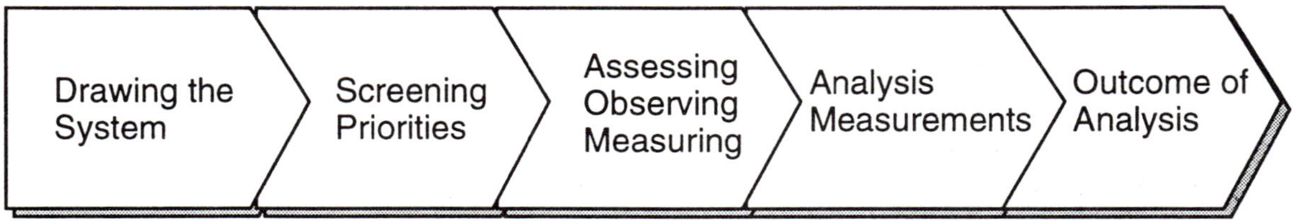

Figure 5.4 Sulzer's process flow for 'understanding the environment'

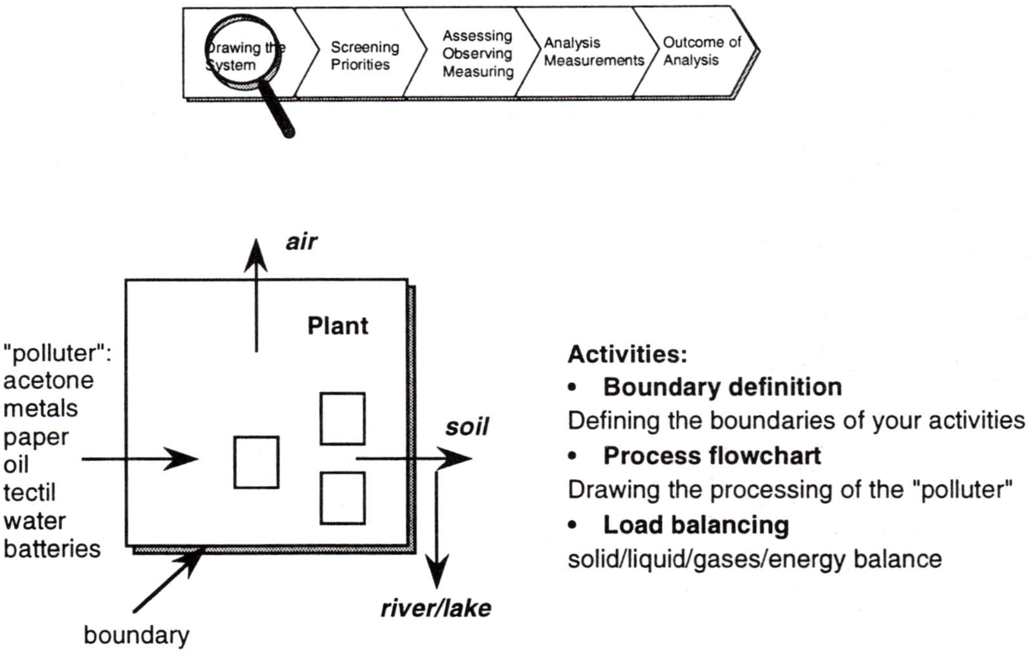

Figure 5.5 Sulzer's sub-process for 'drawing the system'

[18] Entec is a consultancy firm based in the UK that specialises in several environmental areas such as waste management, environmental engineering, environmental managment, planning and development, integrated risk management, contaminated land remediation, to name a few. Entec became part of the Lyonnaisse des Eau Group in late 1995.

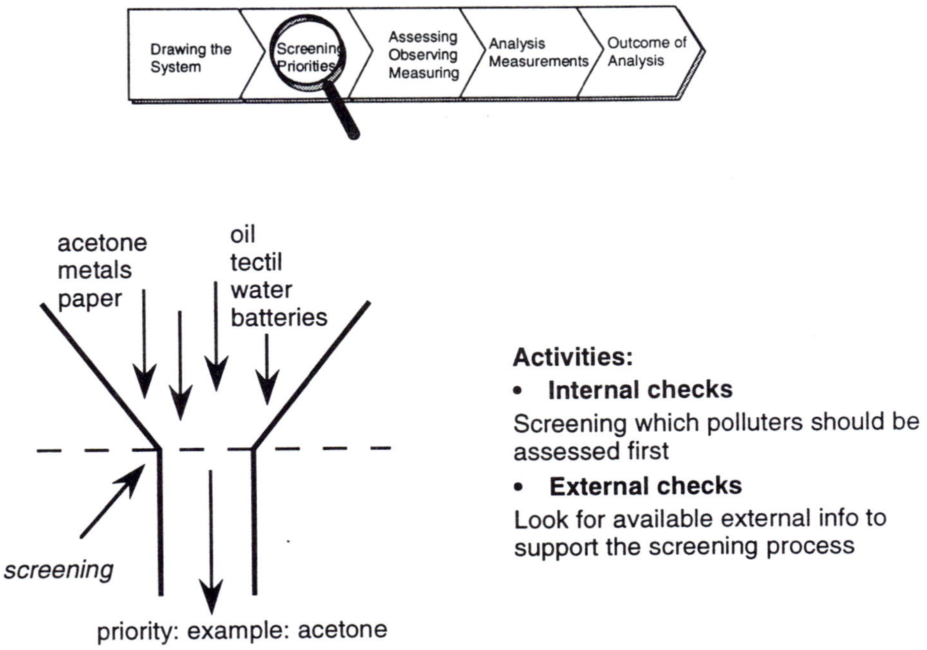

Figure 5.6 Sulzer's sub-process for 'screening priorities'

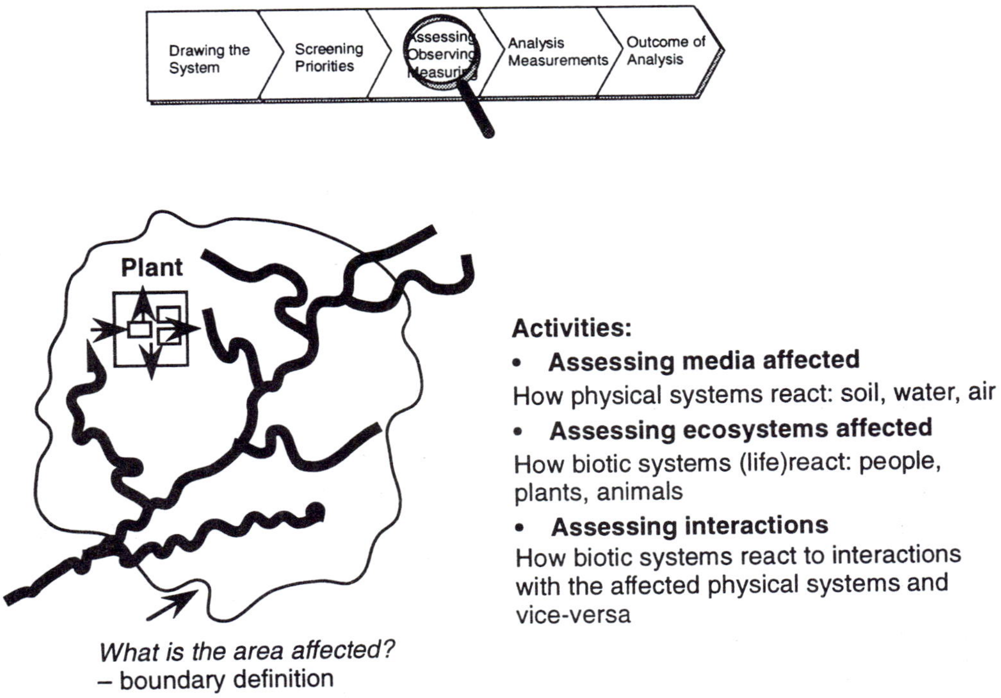

Figure 5.7 Sulzer's sub-process for 'assessing, observing, and measuring'

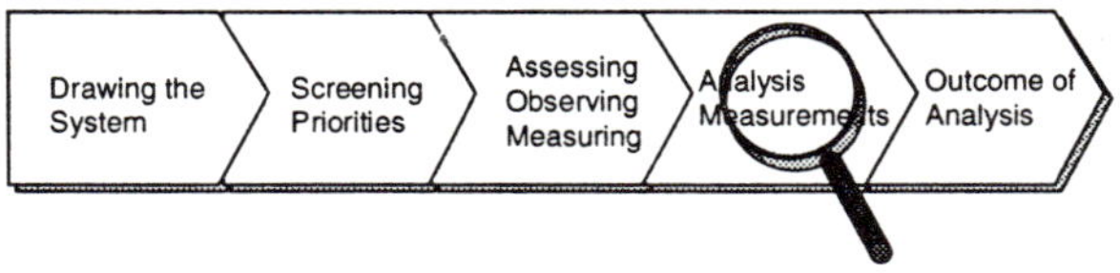

Activities:

- **Mapping**

Drawing the physical and biotic components affected around the area of the plant

- **Modelling/statistics**

Establishing mathematical models defining the interactions among those components

Figure 5.8 Sulzer's sub-process for 'analysing measurements'

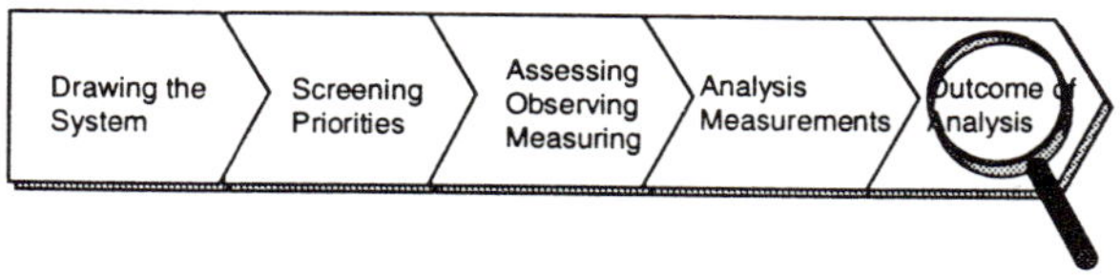

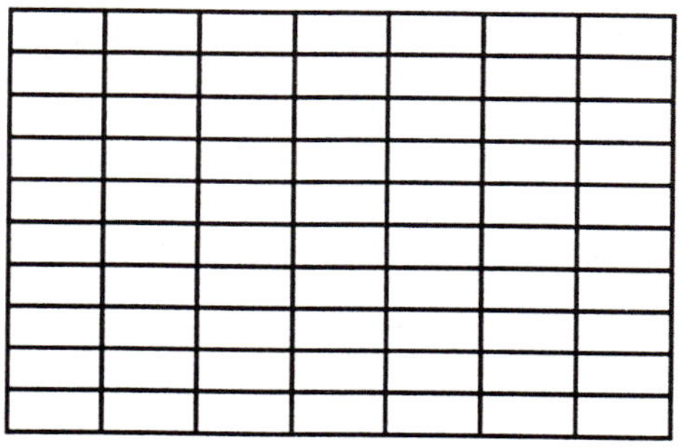

Activities:

- **Analysis**

Defining sustainable limits of IWO's for the plant

Figure 5.9 Sulzer's sub-process for 'outcome of the analysis'

The sub-processes of the 'understanding the environment' process are shown in Figures 5.5 through 5.9. For each of the sub-processes, the team developed the various steps required to apply the sub-processes to the plant issue of acetone usage. At this stage, the work became very intense for MSB. The line manager's role was to help the other team members understand the detailed sub-processes - as well as to evaluate the efficiency of the sub-processes - using the MSB plant as a test site (see Figures 5.5 through 5.9). When this re-engineering or engineering phase was completed and the implications of the system were understood by the entire team, they were ready to begin benchmarking.

Doing some homework The Sulzer team was facing several constraints regarding finishing the environmental benchmarking study. They needed results within three months of their visit to Entec. This meant that the benchmarking partners needed to be located close to Winterthur. As well, because the system being benchmarked was a 'greenfield' system, they wanted the point of view of three more diverse benchmarking partners:

* A best practice company to ensure that their process was practical and implementable;

* An expert in the field of environmental measurements, i.e., a technical academic institution - to ensure that the team was developing something new that would help them develop a competitive advantage;

* An economist or risk management consultant to check the soundness of the process from a macro perspective.

After collecting many articles, pursuing business contacts, reading journals and attending conferences, the team decided to use the following organisations to complete its benchmarking exercise.

<u>A chemical company with noted excellence for its environmental program</u>
- a world-renowned best practice proactive company, accustomed to looking for new solutions
- contact could be easily made from a high level to a high level
- located in Switzerland

<u>An academic institution</u>
- one of the top Swiss technical universities
- currently has PhD students working in the area of modelling the interactions of substances in the eco-system
- noted for its state-of-the-art research in the environmental field

<u>An engineering consulting company</u>
- consulting engineers who specialise in the development of models and tools in the area of risk management
- contact easily made from a high level to a high level

Preparing to make the visit Prior to making the benchmarking visits, the team sent the respective organisations a description of the environmental benchmarking study that Sulzer was conducting, copies of the 'understanding the environment' process and a detailed breakdown of the sub-processes as Sulzer understood them. The partners expressed considerable excitement, as they were thinking along similar lines, and there was a very positive feeling about the exchange of information that would transpire during the visits - the *quid pro quo* being the knowledge transfer between the organisations. The visit itself and the ensuing action plan will be described in the next chapter, after a clear description of 'Phase 3: Action, Conducting visits and Developing Implementation Plans.'

Chapter Six

PHASE 3: ACTION
Conducting Visits and Developing Implementation Plans

Phase 3 of the environmental benchmarking process is comprised of four separate steps that require action from the team members. As the benchmarking companies and site preparations were identified in Phase 2, the team is now ready to benchmark. The intent of this chapter is to answer the questions directed at making the action phase successful.

* What is the most productive way to conduct benchmarking visits and how can the team derive maximum value?

* How does the team clearly identify the gaps?

* How should the gaps be translated into a plan for implementation?

* How should the results of the study be communicated?

* How did it work at Sulzer?

Making the Visit

Gathering information In order to derive optimum value from the benchmarking site visit, each team member needs to gather specific detailed information. One approach that works well is to divide the questions into blocks which are thereafter assigned to particular team members. Each person is responsible for applying the analysis to his or her block of questions, as well as for taking notes while other team members are asking questions. It is good practice to have more than one team member taking notes simultaneously, to ensure that all information is recorded. Having one team member in charge of creating an appraisal of the general atmosphere of the site is also worthwhile, including the company's environmental culture - and an assessment of the company's EEP status. Observing the attitude of the employees, the physical appearance of the site, communications posted on information boards, in addition to the formal questioning, can greatly assist the information gathering process, as well as the subsequent deployment of new ideas.

Fundamentally, the information gathering activity needs to focus on the 'what', the 'why' and the 'how'. Again, these terms refer to measures of environmental effectiveness, the underlying processes and systems that create good results, and the management/culture of the company that fosters ever improving environmental practices. This step needs to be done in a thoroughly detailed way. It is essential to collect the data necessary to measure gaps, e.g., the amount of fossil fuel energy used per unit of production, the detailed operating process that enables the company to use paint with 50 percent fewer air-borne pollutants, the people involved in making purchasing decisions, their specific environmental training and education, whether or not the emissions data are computerised, frequency of data collection, the major source of innovative ideas, the systems used for support, etc.

Each member of the team needs to practice *active listening*; fundamentally, they must understand the 'what', 'why', and 'how' well enough to form concrete ideas as to how the best practice concepts can be brought home and integrated into their own company operations. This means interacting during the questioning period in order to clarify responses, and then probing thoroughly so that all relevant details of the process are uncovered. Achieving this objective often requires going beyond the question and answer sessions to seek out other valuable information. A good practice is to begin by sorting through explanations of the 'what', 'why' and 'how'. From this point on, actual observation of physical processes is critical. Often, this review will uncover inconsistencies with earlier information and conclusions, in which case the 'what', 'why' and 'how' should again be noted carefully. Making diagrams is a good way to gain a better understanding of how the benchmarked company actually does the process.

Immediately after the visit, the benchmarking team should meet in order to confirm their respective notes, clarify points of misunderstanding, and determine any gaps in the information. This step should take place on the same day or the following day so that the opinions and impressions are still fresh in everyone's mind. As well, a debriefing session where the team members assess what makes the benchmarked company's process superior to their own - the overall perception of the company's environmental attitude - will reveal each individual's perceptions and highlight the underlying differences (identify the gaps) between the performance in the two organisations.

Giving information Companies that understand benchmarking will expect that any serious benchmarking team will be able to see *some* improvements that might be made in its operations. In most benchmarking situations there is an expected quid pro quo that encourages information sharing from both sides of the benchmarking partnership. As noted in Chapter 3, before Xerox agrees to be benchmarked, they want assurance that they can also gain an idea or two from the requesting company. This quid pro quo might also be in a completely different area - one where the visiting company is regarded as having some particular expertise.

One of the author's experience visiting several Japanese companies provides a good example. After the visit was completed, the individual was asked to identify anything he saw that might be improved. His remarks were carefully noted by the equivalent of a benchmarking team. Several other people in the conference room were taking notes, probing his ideas and asking relevant questions in order to reach their own conclusions about the feasibility and value of his suggestions. Any benchmarking team should expect to be similarly probed by the company they are visiting. It could occur immediately or, in other cases, it might take place later - sometimes during a return visit. The key point is that excellent benchmarking means looking for ways to learn when others visit - as well as gaining something from visiting elsewhere. Slightly different skills are required, but clearly there is *always* an opportunity for complementary learning.

Identifying the Gaps

Gaps must be seen as opportunities In many ways, clearly identifying gaps is one of the most important results of the benchmarking process. Finding gaps leads to goals and action plans, so these gaps need to be very clearly understood by the benchmarking team and accordingly communicated to the rest of the organisation. The team must use the gaps in ways that lead to improvement - with as little criticism of specific individuals as possible.

After the visit, it is essential that the benchmarking team spend the time necessary to state the gaps explicitly. This means reaching a consensus on the 'what' (what are they using in terms of concrete measures?), the 'why' (why do they use this detailed process analysis to derive these measures?), and the 'how' (how does this company achieve its environmental culture?). It is important that the time required to reach these conclusions not be underestimated. They need to be as explicit as possible so that they can be communicated to others; they also need to lead to an action plan for the company - one that is both ambitious and *doable*.

Achieving clarification The benchmarking team needs to reach this consensus as soon as possible after the visit, while the ideas and information are fresh in everyone's mind. In many cases, the team realises that it needs a few key points clarified. The team should push their work towards a concrete conclusion, including the development of an action plan well underway, before asking for further clarification. Sometimes, more internal information must be gathered; that is, the gap requires having a better understanding of certain information in the benchmarking team's own company that it still does not possess. Usually, the company that was visited can provide any additional information required. In most cases, such requests for information should come from only one member of the benchmarking team. Moreover, in order to improve the benchmarking practice continuously, the team members should always ask themselves why they had not gathered that information when they were on site.

Achieving consensus It is worth re-emphasising the point about criticism. A blame-free perspective toward current environmental practices is essential to making major improvements. The goals and action program formulated by the benchmarking team need to be ambitious - a 50% reduction is a reasonable goal. But, this should never be interpreted by the present process owners as an indictment of their practices.

Improvement targets should be high enough to make a real difference in the environmental results. The biggest challenge at this point frequently is in determining the degree to which the current processes need to be changed. But this is not always the case. The focus might also need to be on the 'how', i.e., the environmental improvement culture that prevails at the company - does it need to be improved? For example, it might be necessary to increase or change the form of internal communications, or to involve a cross-functional group in developing the system; top management may need to have more dialogue with external stakeholders; or a large capital expenditure may be required for new technology that will lower specific emissions. The key point here is that in order for the company to close significant gaps, upper management involvement is critical at this point. The team needs to communicate the facts discovered during the benchmarking visits, the resultant conclusions as to the key gaps, and to develop alternative plans for action. All of these efforts need to be endorsed by a senior management group that has a *genuine* belief in the environmental imperative: they *do* see the achievement of best environmental practices as creating a distinctive competitive advantage.

Developing the Implementation Plan

The implementation team After the team and a senior management group have evaluated the alternatives and conclusions have been reached, the company is ready to develop a detailed implementation plan. This task is typically accomplished by a team, but not necessarily the environmental benchmarking team. One reason is that the benchmarking team may have only two of the final process owners. It is best to have more of the current process owners fully participate in devising the final action plan. What is required is a 'baton pass' from the benchmarking team to the implementation team, where the 'what', 'why' and 'how' are clearly communicated, and where certain benchmarking team members play a prominent role on the implementation team.

Timing is important as the level of commitment to make the changes is highest just after the benchmarking site visit. Including more of the process owners may initially slow down the planning, but the results are usually much easier to implement and more acceptable to the end users. Thus, keeping the momentum high and working for timely results is critical to successful benchmarking.

The implementation team needs to prioritise which aspects of the process to change first. Can the entire process be changed simultaneously, must we proceed sequentially, or does one part need to be altered before another so that interruptions are kept at a minimum? To what extent does the change project embrace the 'how' versus only the 'what'? When a culture change is clearly a key part of the project, it may be wise to select a sequence that involves individuals who are more receptive to new environmental ideas. Getting early wins and adopters gives positive endorsement to the new environmental imperative. The overall environmental culture of the company must be kept in mind when developing the change implementation plan. Some managers should be expected to not understand or believe in the new approach of regarding the environment as a competitive opportunity and may strongly resist new processes and ideas. In such cases, being able to attain early wins which serve as examples of success are useful. How the implementation team members are chosen is critical - for implementation to proceed, the people on the team must be good change agents and be respected by their peers as well as by upper management. This was the reason that the Sulzer team included the production manager assistant who was to implement the change at a very early stage in the environmental benchmarking process.

Identifying the results A cost/benefit analysis can be useful in gaining acceptance of the implementation plan. Showing the advantages of meeting or exceeding the level achieved by the benchmarked company in financial terms is good support and difficult to resist. Conducting a cost/benefit analysis may be difficult if the appropriate information is not gathered during the benchmarking visit.

However, an environmental benchmarking study should be expected to result in an improvement project that needs to be measured in terms of new variables. The standard cost/benefit analysis needs to be enhanced - where environmental variables are codified, set as goals, and monitored for success. At the same time, the entire company needs to understand *why* these goals are consistent with competitive viability. A successful implementation plan is one that is doable and for which progress toward the final goal can be readily measured at various intermediate points. Such a plan should include, but not be limited to, the following:

* Specified, codified environmental objectives - *stretch* objectives,
* A timeline for the change with milestones for the various steps in the plan,
* A resources required for the change along with a list of all departments or divisions that will be affected,
* The specifics of the change and who will be involved,
* Expected costs that will be incurred,

* Final objective/target at each milestone,
* A system for measuring successful completion at the end of the implementation process and, more importantly, at each milestone.

Communicating the Results

Fostering the 'how' Each environmental benchmarking project has the potential to make a strong influence on the company's environmental culture - and the extent to which the environmental imperative becomes a reality. Good results need to be well communicated, and poor results should not be ignored either. The reasons for not reaching environmental benchmarking study results should be reviewed and the appropriate lessons learned. Internal communication of environmental results fosters the commitment toward achieving competitive advantage, and puts further environmental efforts in the right focus. External communication of environmental achievements is a somewhat different matter. Some companies do indeed publicise their results, but others adopt a relatively low profile, being wary of setting themselves up unnecessarily for any accidental problems that might occur.

For the implementation team - as well as for the benchmarking team - it can be very useful to celebrate the achievement of actual results, i.e., acknowledging that their joint efforts have indeed borne fruit. This makes the next project that much easier to undertake, with perhaps even greater stretch objectives.

There is also an ongoing need for continuous improvement. The implementation team, after achieving the desired results, should never stop looking for ways to accomplish the task even better. Measurements should continue, new goals should be set, and perhaps new environmental objectives undertaken.

Communicating with the benchmarking partners It is also a good idea to take the time to communicate the results back to the companies in the benchmarking study. They will be interested, they will appreciate the results, and they might even want to initiate some form of follow-up work.

Sulzer's Implementation Plan

Chapter 5 described three environmental benchmarking partners for Sulzer: a multinational chemical company; an academic institution; and a risk management consultant. The Sulzer team sent three people on each visit. To ensure the best use of time in the interviews, a description of the research that had been conducted beforehand and a copy of the proposed process for 'understanding the environment' (as shown in Figure 6.1) was sent in advance.

All three Sulzer team members were involved in asking questions and taking notes. Immediately after the visit was completed, the team met to clarify any points with one team member taking responsibility for writing up the 'minutes'. Each benchmarking partner supplied the team with the necessary written back-up material in order to enhance the visit and provide facts for preparing the minutes. The minutes were then distributed to each team member, and another meeting was held to discuss the implications. During this session, the team reassessed their position in the process, what had they learned, and any new ideas that could be used at the next benchmarking visit. As the team's philosophy was 'time is essential', the three visits - to gather information, clarify the results and complete the visits - took place within the span of three weeks.

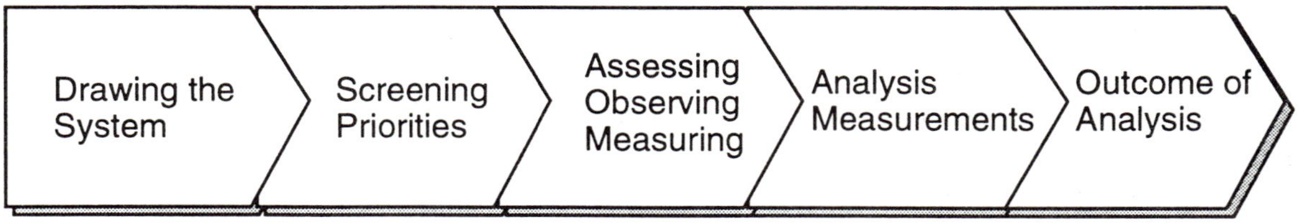

Figure 6.1 Sulzer's process flow for 'understanding the environment'

At the conclusion of the three visits, the team met to compare any diversity amongst their views of the proposed process to understand the environment such that sustainable levels of emissions could be set. The main conclusions were:

* Even though the Sulzer team purposely sought three different perspectives - business, academic, and risk management, they discovered many commonalties in the approaches currently being considered by all three parties.

* The most striking common thread was that all three benchmarking partners confirmed that the process Sulzer had designed was the right process to achieve its goal of setting sustainable limits of waste and emissions, but that this ideal goal was currently not achievable - which forced the team to reassess its original objective - focusing on what are currently 'realistic' limits to waste/emissions.

The key learning points gained from each of the benchmarking partners were:

* The chemical company had established a series of detailed targets and action plans to meet all these targets. It had an in-depth system for assessing Inputs, Waste, and Outputs (IWOs) as well as one for prioritising targets based on the type of material and type of waste produced. Details were shared in written form, which gave Sulzer a set of criteria to be used in identifying the gaps and improving their own proposed process.

* The academic institution used a system that came closest to the sustainable limits ideal. It concentrated on the process-modelling portion of the process, and the resultant physical and chemical interactions that would occur in nature when waste streams were introduced. The system was advanced enough to model the amount of waste/emission, the duration of the polluting period, and the limit at which nature would be altered.

* The engineering consulting company had a process that was the closest to Sulzer's model for understanding the environment. For the complex environmental impact interactions, this company had developed a useful procedure based on risk assessment criteria. The team discussed every step in the procedure, the perceived benefits of using it, and the detailed results. This evaluation allowed the Sulzer team to identify key gaps in their own model for understanding the environment. Additionally, Sulzer learned a great deal about risk assessment in setting environmental performance targets and improvement criteria.

A fourth informal benchmarking exercise was made with a company in the aluminium industry. For this company's facility the environmental impact was physically visible and relatively easy to assess. Moreover, this company was able to develop tangible improvement targets and measures based on the renewal of nature where its smelters were located. Nevertheless, the benchmark was valuable in several ways. The most notable was with the 'how' dimension of benchmarking, i.e., the culture established in the firm for understanding the environment. This company's method of involving all employees in the planning of improvements and the underlying principles of its process for understanding the environment confirmed Sulzer's overall approach - and added a few specific new ideas.

Sulzer's findings and conclusions The team's final conclusion was that Sulzer was on the right track, but that the tools were not yet available to tackle the issue at the level they had established in their model. To make use of available tools and apply its model process, Sulzer would have to adapt it to the current available level of knowledge.

The team's evaluation of the best practice approach to developing a process for understanding the environment led them to the following three conclusions:

* In developing a realistic approach for understanding environmental targets and measurements, there is no uniform approach for establishing boundaries; current best practice is necessarily based on a case by case assessment and evaluation of 'damage'. At the same time, examining these cases is a useful starting point for any new assessment activity.

* Techniques are available for screening, assessing and analysing certain environmental impacts, but they tend to be based on health/safety criteria and on standard physical/chemical properties of substances. The assessment is concentrated on 'levels of damage' to individual components (plants, animals, people, water, etc.), and analysis of environmental interactions *among* the components is in its infancy. However, there are some techniques that can establish the cause and effect relationship of certain 'pollutants' on several discrete components of the eco-system. Most of these models are based on the toxicity levels of well-known substances and consider the impact on equally well-known components.

* Sulzer's model - or that of any other leading-edge company - can only be considered as a set of tentative guidelines for understanding the environment. It appears that defining true 'sustainable limits' is currently beyond man's capabilities. Since we do not know what the absolute sustainable limits are, we need to work *towards* sustainability by reducing the present levels of resource consumption and waste generation.

Sulzer's action plan Sulzer learned - through benchmarking the process of 'understanding the environment' - that in order to improve its overall environmental performance, it needed to inform and influence the actions of people in its business chain. Thus, the team developed the model shown in Figure 6.2. According to these curves, Sulzer's main impact on the environment occurs when its products (industrial goods) are being used rather than during their production. Thus the team's action plan needed to include both a medium and a long-term perspective where Sulzer would develop a 'design for environment' competence. The company concluded that enhancing the environmental attributes of its products at the design stage would offer the best potential for critical improvement - as opposed to enhancing the manufacturing processes themselves. Accomplishing this objective would require that the research development group focus more closely on the environment, by acquiring the right skills and technical resources to create best-in-class designs for environmental practices. Designing for the environment would need to become a main strategic thrust if Sulzer wished to maintain its competitive advantage in the sectors it was serving.

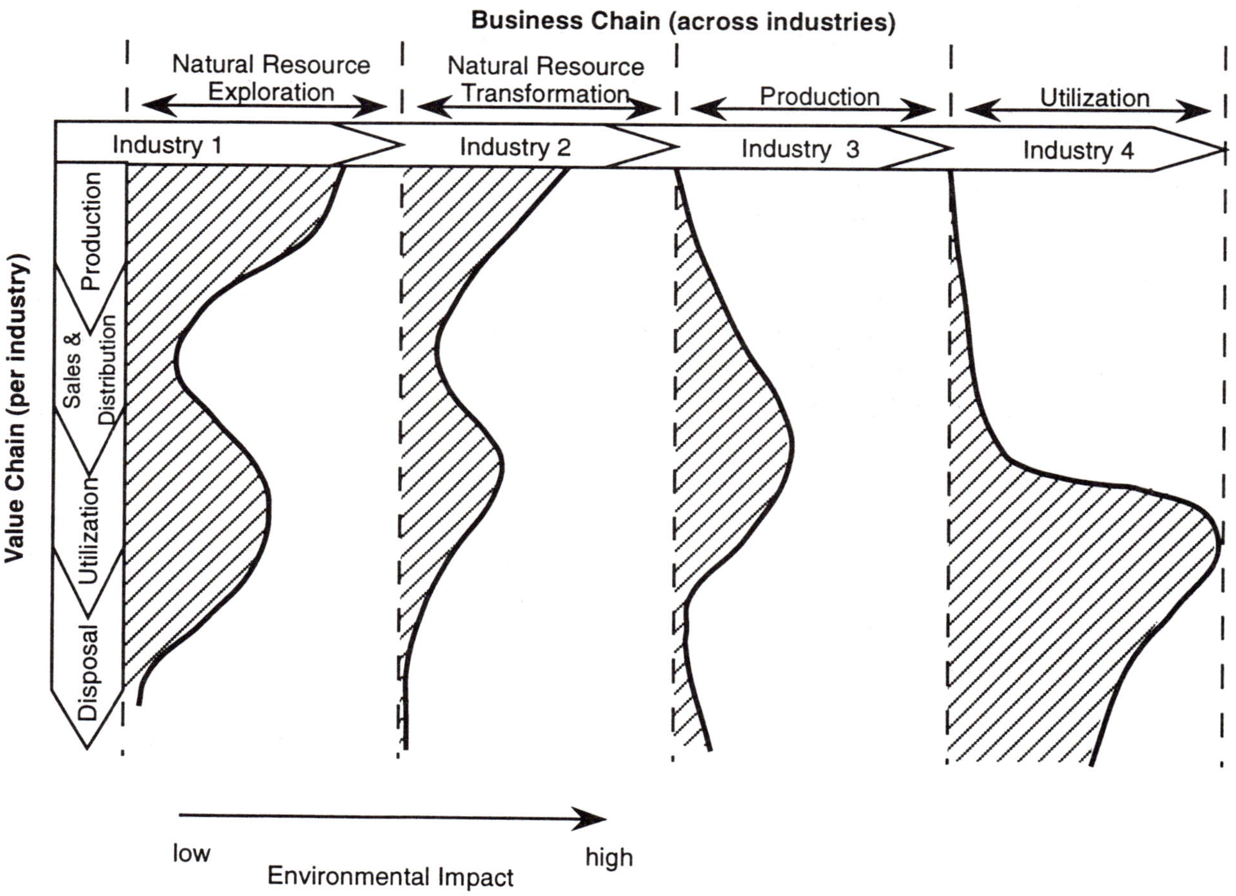

Figure 6.2 Environmental impact of products throughout their life cycle

The action plan proposed to the board was to develop a comprehensive environmental management system. The team presented evidence that supported the need; they had developed a realistic model on which to base such a system; and they had proposed a step-by-step approach for working toward attaining sustainable waste/emission limits, i.e., optimising the consumption of resources and minimising waste through setting stretch targets, measuring improvements, and extending these targets further. The team also felt that they could leverage Sulzer's TQM system if the employees adopted an EMS (Environmental Management System).

Other actions that were implemented included:

* *Improved training and education on environmental issues* In 1995, Sulzer held its first workshop on the environment, entitled 'Environmental Protection - Challenges and Chances'. The purpose of the workshop was to provide the management teams of Sulzer's different product divisions with a forum to discuss the following issues:

  - How do your customers respond to the environmental challenges that confront them?
  - Which demands will they have to cope with in the future?
  - How will these affect Sulzer as a supplier?
  - What are Sulzer's chances?

  The 60 managers attending the workshop left with firm plans to have subsequent interactions of this type in the future. The workshop was also used to bridge the gap between the knowledge gained by the benchmarking team and the managers of the product divisions. As one of the team members remarked, 'There is a lot of knowledge in the divisions, but we need to develop a cross-sharing culture so that the knowledge is leveraged and used where it is most needed.'

* *Development of a new group called 'Eco-systems' within Sulzer's corporate R&D* 'Eco-systems' will serve Sulzer's product divisions as well as outside customers.

The benchmarking team felt that their 'journey' had not only been productive from an environmental perspective but useful in enhancing Sulzer's competitiveness: they also believed that their own competency had been sharpened by the experience. They were now more environmentally conscious, they were more aware of how to assess excellent environmental practice, and the process of benchmarking itself would contribute to the way each of them intended to work in the future.

Chapter Seven

PHASE 4: CONTINUOUS IMPROVEMENT

The concepts of benchmarking need to be seen within the context of continuous improvement. That is, there is no such thing as the ultimate best practice; moreover, the environmental measures of effectiveness are themselves continually being redefined. Chapter 7 explores these ideas in terms of the following two questions:

* How does a company develop environmental benchmarking within a continuous improvement philosophy?

* How are leading-edge companies applying benchmarking concepts more widely?

Benchmarking within a Continuous Improvement Context

Environmental benchmarking, like any other form of benchmarking, needs to be seen in a spirit of continuous improvement. It will not be done only once, the objectives will change, and the particular topical focus will also evolve. This is nicely illustrated by returning to our environmental benchmarking diagnostic as shown in Figure 7.1. The diagnostic facilitates a journey from the discontinuities and changes in expectations a company faces to the strategic responses that will satisfy the company's strategic intent. Thereafter, we develop the bottom half or infrastructure part of the model through competency mapping, which yields the most important processes and systems that are needed to build the core competency identified. These are the primary candidates for benchmarking. But, after benchmarking and improvement efforts have been completed in these infrastructure areas, the obvious question is: What is next?

One answer is to begin again, to take a fresh look at the discontinuities and expectations - particularly if there has been some important change that represents a new challenge. For example, Sulzer might find that some new regulation has an important impact on environmental incineration, which needs to be included in the way the company designs its products. This might, in turn, lead to a need for some additional competencies, such as new ways to position its products and services or some new technical requirement for products that necessitates making an alliance with another company. In a similar fashion, certain important municipal customers might find themselves in a position where some form of equipment rental is preferable to purchasing that equipment outright. This would imply a need for Sulzer to develop new financing competencies, within its underlying infrastructure.

Other changes in the top half of the model might include a shift in the basic strategy or strategic intent of the business. Thus, one company was largely in the business of certification, particularly for ISO 9000 quality. They saw the emergence of environmental certification as a distinct business opportunity and incorporated this belief into the company's overall mission. This, in turn, led to a need for several action programs and related competencies - including new information systems capabilities that focus on environmental traceability and reporting. The key point here was that this firm took a proactive approach to the environmental issue and the related benchmarking and improvement activity. A potential business opportunity was seen and the company believed that being first and foremost with the competencies to support this opportunity was indeed strategically important.

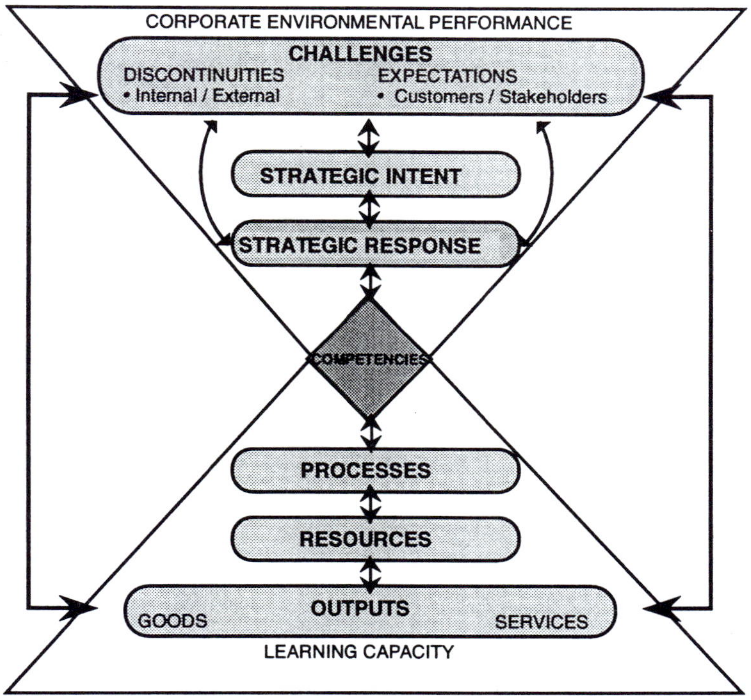

Figure 7.1 Environmental business transformation diagnostic

Still another firm needed a fresh approach to environmental benchmarking because the employee stakeholders became very concerned about a possible carcinogenic problem with the present processes. In this case, there was more at stake than just dealing with the problem at hand. The company saw a clear need for having comprehensive environmental auditing and training programs. In this case, the benchmarking focused explicitly on systems that would support employee knowledge about the environmental issues associated with particular industrial processes.

Continuous improvement does not, however, have to entail a comprehensive review of the environmental strategy. The improvements can be focused entirely within the existing set of perceived competency needs. The process of environmental competency mapping develops a tree-like structure for infrastructure requirements, with a resultant prioritisation of process improvements. But, once a particular process has been improved, it is natural that the company will turn to those processes that rank next in perceived importance. An example for a company in the chemical industry was a perceived need for environmental information systems competency. This, in turn, led them to realise that a common reporting system with common data elements and approaches was critical. They benchmarked the systems being used by several other companies and then designed their own information system. But, after it was in place, the firm perceived the need for having information systems that would support consistent methods for performing life-cycle assessments. To achieve this goal, they would need to benchmark with firms performing life-cycle assessments and find those that used a common methodology to obtain consistent results in varying problem situations.

Another example was when a company making plastic products began focusing its environmental activities on developing infrastructure that could answer *any* customer's questions about the environment and do so with maximum assurance of accuracy in a short time frame. This objective led them to benchmark with companies in completely different industries and to begin by developing the necessary information systems to support their customers' requests. But, thereafter, they realised that additional competencies were required on the part of all employees to utilise this infrastructure - not only on a reactive basis to respond to requests - but also proactively to clearly demonstrate the firm's concern about environmental issues.

A final issue in 'continuous improvement' in the environmental field is deciding how the improvement will be carried out and how setting the direction for this activity will be established. Leading-edge firms do note see continuous improvement as an accidental response to problems or as being directed in some random fashion. Continuous improvement is much too important to leave to chance. Even more fundamentally, continuous improvement must be seen as critical to success - the speed of improvement will separate the winners from the losers. The central tenant of this book - *the environmental imperative* - is that those firms which carry out environmentally-sound business development will dominate their industries.

The imperative now needs to be seen in the context of a *process* for continuous improvement. Environmental benchmarking must take place at a rapid pace. It needs to be undertaken with the same sense of urgency a company would apply to any other critical initiative. It is not an add-on to the business that will merely be done in spare time or when there is no other pressing work to do. No, the environmental imperative demands progress on environmental issues consistent with progress in all other relevant business concerns - not because we are good citizens, but because we are smart business people. This means that the environmental agenda needs to be determined deliberately, monitored for efficiency, and assessed for effectiveness. That is, any given set of environmental objectives needs to be assessed continually to ensure that progress is being achieved within the targeted time frame. It is perhaps even more important to assess the organisation's overall environmental goals, how they are being set, and how they are being updated over time.

During a presentation on a complex information system to the board of directors of a large corporation, one board member asked: 'If you were in my position, what would you worry about?' In the environmental context, our answer would be that the firm is not achieving optimal progress in terms of the environmental imperative. Concerned senior executives should worry that their continuous improvement process for the environment is not robust enough - both in terms of achieving the stated objectives and determining the right objectives in the first place.

A new kind of benchmarking is implied in this concern. Which companies *do* have excellent continuous improvement activities for the environment? How are they designed and monitored? What competencies and infrastructure support these activities? How can managers go about examining the continuous improvement activity in other companies - and then use the knowledge gained to foster their own continuous improvement?

Applying Benchmarking Concepts More Widely

Benchmarking in its widest sense seeks to learn from the experiences of others. Although we are primarily interested in the practice of environmental processes per se, there are other types of learning that can be shared. The general concept of 'practice sharing' - even beyond the practice of processes - needs to be fostered in several ways.

Practice sharing Practice sharing seems to be either totally lacking or at least inadequate in many firms. We see company after company where sister divisions and facilities do not share best practice information. Many reasons account for this failure, but the primary one appears to be the competition between units, fostered by making comparisons. The most detrimental ones are financial comparisons based on accounting and other information. In far too many situations, when a plant or division is rewarded for excellent performance, it is reluctant to share the reasons for this performance with its sister divisions/facilities - as they are seen as competitors.

Companies need to escape from this kind of thinking. They should also presume that every division or facility can do *something* in an excellent way and, if the entire company could adopt the best features of *every* facility, it could have a formidable competitive weapon. This problem is similar to the 'feel good' trip issue outlined in Chapter 5: many business units dislike admitting that they can learn something from their sister units. The second most common inhibitor to practice sharing is a decentralised structure that is not adequately supported with communication linkages between business units/facilities. In order that information and best practices be shared and readily available, a systematic approach to knowledge and information transfer must be supported.

Practice sharing also needs to be seen as a two-way learning experience, not just as giving away one's *own* good ideas. When a group or firm shows another one its business practices, it should expect that it will also learn from the experience. If the people from the visiting company have done their homework properly, have gone through their own business process re-engineering, and followed correct benchmarking procedures, then they should be able to make a critical assessment of what they observe. Perhaps they will be the major beneficiary, but they need not be the only one. After the team has looked critically at the process they came to see, they should be asked to recommend areas that could be improved - based on their observations. After all, the members of the team are, by definition, well qualified and should therefore be able to find a suggestion or two for improvement - based on *their* experience.

A more general view of practice sharing is also useful in environmental benchmarking. Although the most straightforward comparisons are based on the practice of processes, firms can also share more imprecise concepts and even philosophies about the environment. In the MIBE project, we have found that firms are usually interested, for example, in what others think about employee education, databases, measurement philosophies, experiences with environmental certification programs, and approaches to learning from environmental pressure groups.

Forms of environmental response In Chapter 2, we described an evolution that seems to occur regularly in the environmental field. Environmental problems are often attacked first by governmental units that use command and control approaches to regulate the behaviour of businesses and other actors. At some point these approaches are likely to yield to market-based responses, and firms that are in the best position to make environmental differences are encouraged to establish marketplace activities. Eventually, industry groups and farsighted firms are the ones to take responsibility for establishing the environmental agenda.

The key point here is that the basic type of practice sharing will change completely under different conditions for environmental improvement. Firms should learn to recognise the variations and the natural evolution in these patterns, and then the need for shaping their responses, their practices and benchmarking comparisons accordingly.

Another way of thinking about different forms of environmental response is shown in Figure 7.2. Here, we see a hierarchy of responses, and firms may well be using several different ones at the same time. Starting at the left-hand side of Figure 7.2, the response to environmental issues is labelled 'fighting'. Next, we have 'complying', then 'partnering', 'anticipating', and finally 'setting the agenda'. One might disagree with the use of these terms, but the idea is clear: firms respond differently to environmental initiatives depending on their circumstances and pressures. There will also be differences in how firms handle practice sharing.

Figure 7.2 Evolution of company environmental response

Creating environmental challenges Most environmental benchmarking studies deal with a set of reasonably well understood environmental challenges. However, the way that firms respond to these challenges is not the same at different points in time. A question that often arises is whether it is good practice to be 'ahead' in terms of the progression shown in Figure 7.2. That is, perhaps while others are fighting a new set of environmental rules, leading-edge firms have moved toward compliance. As noted in Chapter 2, when the EPA first issued its regulation for auto emissions, American firms fought hard against these regulations. They hired lawyers to overturn the regulation - while Japanese firms hired engineers to help them meet the regulation - giving them a competitive advantage in the North American market.

Carrying this concept a step further is one of the key points in our transformation diagnostic. Instead of just responding to the environmental challenges, the means to achieving true competitive advantage may be by finding ways to develop customer and other stakeholder expectations within one's particular industry and then *setting* the agenda or *creating* the environmental discontinuities.

The environmental benchmarking process New challenges and potential extensions to the environmental agenda do not need to be seen as threats or problems. They are, in fact, challenges and opportunities. Those firms that find effective ways to anticipate the developments - or even influence them, rather than prevent them - will be in a better position to transform their organisations so as to profit from these developments.

Again, the way to accomplish this goal is to understand the environmental benchmarking process thoroughly. Connect the challenges to the firm's strategic intent, to the strategic response, to the new core competency requirements, to the infrastructure needed to acquire these competencies, and finally to those processes and systems that are most critical for achieving distinctive competencies.

Moreover, one should explicitly understand the evolution from 'fighting' new environmental challenges to the end result of meeting and even seeking these challenges in a proactive fashion. Finally, again, it is useful to apply the environmental toolkit whenever possible: total quality environmental management, environmental, health and safety, cross-functional management teams. The list is endless; therefore, management must decide where and when the environmental toolkit can most effectively be applied to develop a competitive advantage.

Companies with best practise in environmental benchmarking have moved according to new knowledge, available technology and economic conditions. In this sense, there has been a well-shared need to keep working towards better environmental performance in a continuous fashion. The process of continuous improvement must be considered as the basic philosophy of, and approach to, environmental benchmarking. In the next chapter, we give a more detailed description of how environmental standards and management tools can be used to foster this notion of continuous environmental improvement.

Chapter Eight

ENVIRONMENTAL STANDARDS AND MANAGEMENT TOOLS

The preceding chapters have described environmental benchmarking as a key means to achieving the environmental imperative. Several complementary environmental models and concepts, that can be used in conjunction with environmental benchmarking, are also available. This chapter is devoted to a general description of the most important ones. Additionally, the chapter describes how these standards have been set and how they might evolve in the future. The intention is to describe basic principles and to make the reader aware that these standards and tools will continue to develop as the environmental movement evolves.

The environmental management standards and tools will be considered by answering the following five questions:

* How are regional environmental management standards being set and what is their impact?

* What are the national and international environmental standards and schemes?

* How is life-cycle assessment being applied to environmental issues?

* How are sustainable development business charters being implemented?

* What environmental benchmarking networks can be consulted?

Regional Environmental Management Standards

Leading-edge companies realise that by making a systematic effort to improve environmental performance, they will eventually reduce unnecessary costs, diminish unwanted risks, fulfil their customers' environmental concerns, and develop cleaner, more efficient innovative processes and services. More importantly, these activities will enable them to develop a distinct competitive advantage. In the process of improving the environmental performance of products and services, some companies have not only become more competitive, they have in fact set the environmental standard for the entire industry.

One good example of this situation is found in the automobile industry. In 1972, Volvo and Bosch embarked on a joint venture to develop a device that would reduce the harmful effects of automobile exhaust emissions. The result was the technology for the catalytic converter that ultimately set the standard for the automobile industry. At that time, only the State of California had regulations requiring the use of this technology by all automobile makers selling cars in California. By the 1990s, catalytic converters had become mandatory in approximately 22 industrialised and some developing countries. Clearly, their environmental vision has benefited both companies by making them leaders in their industry - highly profitable ones as well. The key point here is that leading-edge firms not only *comply* with environmental standards - their actions help *establish* the standards.

In most cases, companies do not act alone to determine best environmental practices. However, even when companies have done the critical work (e.g., Volvo/Bosch), there is still an adoptive phase where environmental standards are established. This chapter begins with a description of the most important regional organisations and activities involved in shaping the standard setting process. Over the past few years, several regional standards have been established - such as the European Eco-Management and Audit Scheme (EMAS), Integrated Pollution Prevention and Control (IPPC), and the Parallel Environmental Accords to the North American Free Trade Agreement (NAFTA). Thereafter, we explain the role of national and international organisations in setting standards for the environment. There is one final caveat: this entire field is continually evolving as new (and existing) organisations create new standards and criteria.

Eco-management and Audit Scheme (EMAS) EMAS encourages companies in the industrial sector to participate voluntarily in a European Union scheme that specifies key environmental issues. The main issues/principles to which firms need to comply in order to be EMAS certified include:

* *Initial environmental review* An initial environmental review of each site, which is based on a comprehensive set of variables/issues as established by EMAS, must be conducted.

* *Environmental management system, program, and register of significant effects* EMAS requires that the environmental management system and program apply to all activities at each site being certified. This system needs to follow the detailed specifications of EMAS (included in Annex 1 of EMAS), which states that a register of 'significant' environmental effects be compiled. The environmental effects of past, current and future activities are part of this register. Separate programs are set up to assess specified parameters to deal with the environmental management of projects relating to new developments, and new or modified products, services or processes.

* *Contractors, subcontractors and suppliers* To achieve EMAS certification, firms need to focus on the upstream and downstream activities as well as their own. This means establishing procedures to ensure that suppliers and those acting on the organisation's behalf comply with the company's environmental policies.

* *External communications* Because providing information to the public is seen as an essential element of good environmental management, EMAS requires that the firm's environmental policy to be publicly available, as well as the information necessary to understand the environmental impact of the company's activities.

* *Internal and external audits* Ongoing certification requires that internal audits be carried out at least every three years, with a formal report of each audit cycle being submitted to top management. It also requires that the policies, programs, management system reviews, audit procedures and public environmental statements comply with the European Union regulations. Moreover, an independent, external verifier must validate the 'environmental statements', which in fact lead to improving the processes that affect the environment.

Integrated Pollution Prevention and Control (IPPC) One of the most recent pieces of environmental legislation in the European Union is the IPPC. Its objective is to control emissions from large - especially polluting - industrial plants inside the EU. The basic objective of the IPPC is to take an integrated approach to licensing new and existing industrial installations. Its key aim is to ensure a holistic approach to pollution control, by ensuring that industrial plants are regulated by one license that limits emissions into the air, water and soil - rather than separate ones for each type of emission. This approach is a first attempt to recognise the interactions between various types of emissions and to address the 'overall load' on the eco-system. By eliminating the need to make separate calculations, i.e., where water pollution effects on soil are unaccounted for, etc., firms are in a better position to consider emissions tradeoffs.

For some businesses, the IPPC is the most significant environmental legislation passed in recent years. It will result in some clear winners and losers in the competitive game. The new licensing rules apply to a wide range of industries - from energy, steel and chemicals to food processing, factory farming, and the pulp and paper sector. Some EU member states - including France, Ireland, the UK, and Denmark - have previously taken a similar integrated approach to pollution control in their national legislative actions. This implies that the companies in these countries have a head start in complying with the IPPC statutes.

This legislation also stipulates that three years after the directive is approved, no new installation covered by IPPC will be allowed to operate without the IPPC license. The expectation is that this new approach will encourage industry to move away from 'end-of-pipe' pollution control solutions toward concentrating on designing plants and processes in an overall more environmentally-sound manner, from the initial concept to the de-commissioning phase.

North American Free Trade Agreement (NAFTA) In the three North American countries of the NAFTA region, environmental management standards which guide industry and business are part of the agreement in parallel environmental accords. NAFTA does not, however, have environmental management standards as explicit as what the European Union has through EMAS. The primary objective of the NAFTA environmental agreement is to ensure that the parties enforce their domestic environmental laws. Under the agreement, governments of the three countries are required to enforce environmental agreements and respond to individual requests for investigations. The convention's provisions do mandate several environmental conventions - including the International Pact Prohibiting Illegal Trade, the Basel Convention on Hazardous Wastes, and the Montreal Protocol on Substances that Deplete the Ozone Layer. Although the NAFTA countries may set domestic standards at any level they choose, there is a caveat: the countries must seek 'legitimate' environmental objectives. Since the governments of the three NAFTA countries (as well as the individual companies in these countries) are moving towards the use of the international standard - ISO 14000, 'legitimate' environmental standards will become increasingly common. This migration is also driven by firms that wish to export into global markets.

National and International Environmental Management Standards and Schemes

There has been a proliferation of national environmental management standards and schemes. Each one has been influenced by local conditions, and responds to methods and forms that companies in a particular country can adopt more readily. This is happening at a much faster rate than the ability of multinational companies to learn and understand on a country-by-country basis. The risk here is that the environmental management programs developed by companies will have to be modified, possibly substantially, to meet individual national standards.

Dutch initiatives The Netherlands pioneered voluntary agreements on environmental management as early as the mid-1980s. The Dutch government began talks on voluntary agreements with business sectors on compounds or products considered to be environmental threats. Examples of such agreements include the commitment by businesses to limit the amount of mercury in batteries (1985) and the use of cadmium in crates (1988). The Dutch Environment Ministry has 36 such covenants in place, and the Economic Affairs Ministry has reached 23 long-term agreements on energy efficiency improvements since May 1992. The benefits of developing covenants on the environment include:

* They usually support legislation;
* They can bridge the period between formulation and enforcement of regulations;
* They avoid the problems of the command and control approach by governmental agencies;
* They are considered the most efficient instrument in circumstances where regulations are lacking.

Under the covenants, the relevant industry sectors commit themselves to contributing a proportional amount toward accomplishing the national emission reduction goals as set by the Ministry of the Environment. Targets for emission reductions of substances such as heavy metals and compounds (e.g., nitrogen oxide) are set for each industry sector. Individual companies within that sector make a commitment to develop the environmental business plans needed to attain the toughest goal for the industry sector. We believe that those Dutch firms taking the most proactive stance have the greatest opportunity to achieve competitive advantage based on environmental variables.

German initiatives Other countries, such as Germany, are also encouraging specific industry sectors to improve their environmental performance voluntarily. For example, the German Environment Ministry is encouraging electronic manufacturers to agree on a voluntary recycling scheme rather than be confronted with a mandatory ordinance on used electrical and electronic products. The Ministry has taken a similar approach by encouraging the recycling of used cars. In a recent proposal, the automobile industry responded by proposing to take back and recycle abandoned used cars whose last owners cannot be identified. The industry has also pledged to increase the recyclable content of car parts from the current amount of 75 percent per unit weight to 85 percent by the year 2002 and to 90 percent by 2015.

One very successful covenant in Germany has been made by the paper industry: paper mills, printing firms, paper wholesalers, publishing companies and others involved in the industry have committed themselves to producing and distributing only paper-based products with at least 60 percent recycled content. In return, the Ministry has agreed to drop plans to enact a used paper recycling ordinance, which would have mandated participation in a prescribed collection, recycling and distribution program.

British initiatives The United Kingdom is the first country to develop an environmental management standard (British Standard BS 7750), which was developed by the British

Standard Institute. This standard is voluntary and is based on an environmental management system with third-party certification. It is a natural evolution from the earlier British Standard, BS 5750 - which is the basis for much of the requirements in the ISO 9000 standard for quality management systems. BS 7750 has likewise played a strong role in the development of ISO 14000 standard for environmental management systems.

Some companies will use certification to 'prove' that standard procedures for dealing with the environment are in place. For them, certification is regarded as the primary - or only - goal, while others see certification as merely a useful step, i.e., 'before we can improve, it is necessary to clearly document what it is we are going to improve.' The environmental management system outlined in the BS 7750 (or ISO 14000) is generic in that it is designed to suit all business and industry sectors. The standard is mainly concerned with defining clear roles and responsibilities, targets and objectives.

Northumbrian Water Limited,[19] a player in the UK water industry, decided it would be the first water service company in England and Wales to be certified to the environmental standard, BS 7750. Becoming certified was not only to build a good image as a leader in the recently deregulated water industry, it was also viewed as enhancing the competitive objective of achieving truly sustainable development. Management subsequently developed a conceptual framework for sustainable business development based on four key environmental issues: *measurement* of environmental effects such as the consumption of fossil fuels and atmospheric emissions of greenhouse and ozone-depleting gases; *significance* of the environmental effects to include public perception of the effect; *life-cycle analysis* as a technique to be applied in assessing choices between products with competing environmental claims and complex ecological interactions; and *continuous improvement* in the form of publicly reporting environmental goals, targets and progress, and developing effective decision-making tools.

Other national initiatives There has been much variability in national environmental standards and schemes. For example, the French Government has developed a voluntary agreement with industry on energy consumption and carbon dioxide emission reductions. In other countries, there are little or no requirements for such environmental management standards. For example, in the US., neither the Environmental Protection Agency (EPA) nor the Department of Justice have made strong recommendations to companies to carry out environmental auditing programs. On the other hand, the Netherlands and Norway have legislated some environmental standards in fairly general terms.

International standards A key issue in the development of international environmental standards has again been the influence of individual companies. There are two complementary driving forces for this influence. First, companies operating on a global basis do not want to comply with vastly different standards in each geographic region. This would be confusing, expensive and difficult to explain to employees and others. Moreover, firms operating on a global basis are sensitive to accusations that their environmental commitment is different in different places. The second driving force by companies is a desire for a 'level playing field'. That is, good firms *want* to compete on a basis that includes environmental variables - but they want their competitors to play by the same rules.

---

[19] Became part of Lyonnaisse des Eaux in late 1995.

As a result, companies and industries have developed a number of voluntary international environmental management standards and systems for their business operations and related activities. These standards provide companies with common guidelines for organising their environmental staff, establishing and monitoring operating procedures, addressing non-compliance issues, undertaking corrective activities, establishing documentation of these activities, assessing environmental effects of the company's activities, and conducting environmental audits.

Of particular interest in this regard is the International Standards Organisation's ISO 14000. Most companies are familiar with the quality standard, ISO 9000. This standard assures consumers and suppliers that the products and services they have bought have been produced under a standard set of operating processes. ISO 14000 corresponds to ISO 9000, but it concentrates on the environmental aspects of products and services. The concept of ISO 14000 has been to develop a voluntary standard that verifies a company has an environmental management system in place that should prevent or detect environmental problems, independent of the geographical location of company operations. ISO 14000 promotes significant changes in manufacturing and distribution industries worldwide. As with the ISO 9000, many companies believe that ISO 14000 will become accepted both as a marketing tool and for focusing decisions about a firm's environmental commitment.

An analysis of the ISO 14000 certification standards clearly indicates that the BS 7750 served as a model or starting point for the ISO 14000. BS 7750 is much stricter and more precise in its wording and in its demands regarding the actions and systems it expects from companies. In comparison, ISO 14000 is more vague and general - BS 7750 calls for continual improvement of a company's specific environmental performance, whereas ISO 14000 calls for a continual redefinition of the environmental objectives themselves.

It is important to note that ISO 14000 is not about compliance - it is about management practices. The approach is parallel to that for the ISO 9000 quality standards. ISO 9000 certification does not guarantee good quality. It does certify that a firm has developed a set of procedures to follow and that those procedures are followed. That is, there is a standardised approach to designing and making products. Similarly, ISO 14000 does not make statements regarding what is desirable for the environment. Neither does it set environmental goals, performance levels or technology specifications. Rather, the intent is for a firm to have rigorously *defined* its environment policies, goals, and procedures - and thereafter follow them. ISO 14000 certification is focused on a firm having a *working* environmental management system. It should be remembered that a major reason for a company to set up an environmental management system is to explicitly state its environmental objectives, monitor performance, and demonstrate its ability to manage environmental affairs in a sound manner. Achieving these objectives is enhanced through ISO 14000 certification.

When properly managed, ISO 14000 integrates environmental responsibility into everyday corporate decision-making. For example, Robert Clevestine, quality assurance manager at Warman International Inc., a heavy equipment manufacturer from Australia with more than 1,200 employees in 73 countries, says: 'Managers at any company must look at their assets as risks, whether they represent safety, financial, or environmental risks. So they are looking for the best way to manage those risks. And it now appears that ISO 14000 may provide the best model.'[20]

The existence and adoption of ISO 14000 is increasing in importance for companies with international operations. The absence of an international standard can possibly lead to development of informal trade barriers if countries like Germany, France, and Japan develop unique national standards in lieu of a globally accepted international standard. Based on the experience with ISO 9000, the adoption of ISO 14000 is likely to be faster in Europe than in the United States, mainly because European countries have been seeking to develop common standards for many years.

Preliminary cost estimates for ISO 14000 registration, according to experts in the field of certification, is expected to be one-third to one-half as expensive as registering under ISO 9000. For many large companies, this amount will be between $100,000 and $1 million per facility, even if facilities are already ISO 9000 registered. For small and medium-sized companies without ISO 9000 certification, the estimated cost is between $50,000 and $100,000 per facility.

Life-cycle Assessment as a Measurement Tool

In Chapter 3 we showed that the environmental impact of manufacturing and/or service activities in an industry can be benchmarked at any point along its life cycle (refer to Figures 3.4a and b). In leading-edge companies, life-cycle assessment (LCA) techniques have become useful *internal* management tools for leading environmental improvements at every point along the life cycle of a product or service. Basically, life-cycle assessments facilitate comparing the environmental impact of one product or service relative to another. This assessment can be used as a method to measure and compare processes, and it allows companies to focus their attention on critical areas of products/services that they supply in order to maximise the returns on their environmental investment. In the Sulzer example, Figure 6.2 illustrates this tradeoff. Sulzer's processes are relatively 'clean' in that the emissions and wastes from manufacturing are not as harmful to the environment as the amount of electricity that may be consumed while its products are in use. In this case, Sulzer needs to focus on developing better design techniques and technologies that will reduce its products' impact on the environment throughout their entire life cycle.

The use of LCA does have its shortcomings. When it is used as a tool to set industry benchmarks, a problem arises when companies have not completely agreed on how to apply this technique. Should the impact be calculated over the entire life cycle? How should the extraction of raw materials be measured in terms of impact? Does one ton of bauxite make the same impact as one ton of timber? These questions are being heavily debated; scientific measurements of environmental interactions and long-term affects are still in an early stage. However, several companies have adopted the life-cycle assessment approach as a key part of their environmental management system. It has proven to be an effective internal management tool to identify potential areas for improvement, whether these be process related or product input/usage/packaging related. A key problem in LCA is that it is always possible to define

---

[20] House, G., "Raising a Green Standard, ISO 14000 Brings Environmental Risk Management to the Fore", *Industry Week*, Vol. 244, No.14, July 17, 1995.

things in a somewhat arbitrary way - which can lead to conclusions that change from one product or process to another. But LCA does allow a company to evaluate its impact on the environment systematically - by adopting standardised definitions and approaches, even when some of the choices may be arbitrary, after which systematic evaluations can be made.

The view of experts[21] is that LCA should generally be used to identify where improvements can be made to reduce the environmental impact of a product or process in terms of the energy and raw materials used and the waste produced. LCA is a good tool for enhancing understanding, making improvements, controlling environmental quality, and guiding new product development. Using LCA to compare competitive products is often not viable, as environmental impacts are difficult to quantify both financially and ecologically. As well, companies tend not to use the same parameters in calculating LCA, and thus making comparisons between competitors' products or processes becomes difficult. Several practitioners already recognise this problem, and have tried to establish some rules for calculation and utilisation. That is, the application of LCA in a consultative function over time will yield helpful inputs.

In Germany, the Wuppertal Institute has developed the concept of using Material Inputs Per Unit of Service (MIPS)[22] as a practical initial proxy to estimate and compare the ecological impact intensity of goods and services. The aim is to define quantitative, measurable parameters that will evaluate the environmental impact of a product from 'cradle to grave'. By this term we mean evaluating impact with respect to raw material inputs, impact of extraction requirements, processing, construction and operation of plant or office facilities, treatment of emissions or disposal of waste generated, use of the product, and its disposal at the end of its useful life. It is noted here that there is a difference between life-cycle analysis and life-cycle assessment. Life-cycle analysis is a collection of data that indicates how many grams of a certain raw material have been used, whereas life-cycle assessment also evaluates the inventory of data, whether the resources being used are a scarce resource, or if the extraction process causes damaging pollution.

Proctor & Gamble, one of the world's largest consumer goods producers, has used LCA extensively to reduce the life-cycle impact of its products through changes in packaging, as well as to compare its disposable products and other re-usable alternatives (Pampers disposable diapers versus cloth diapers). However disagreement over using LCA in the packaging industry is acute. People are generally confused about the results of LCAs which attempt to compare the environmental impact of packaging options such as paper, glass, metal, PET and others. GE Plastics, a division of General Electric, has produced GECON, a reusable intermediate bulk container that is being positioned as an alternative to disposable packaging. It is not only functionally easier to assemble, more durable, requiring less stacking space and offering splash-proof protection to shipped goods, but it is also fully recyclable as it is made from 100% thermoplastics.

Companies also use LCA in other ways. Cementa Euroc, a Swedish cement producer, has used LCA to gain a better understanding of how the production of cement affects the environment, as well to compare the environmental impact of concrete as a building material

---

21 Comments from IMD's MIBE Workshop on 'What Is Green', April 20 - 21, 1994.

22 Liedtke, Christa, "Evaluating Products According to Material Intensity per Service (MIPS)" presentation at IMD's MIBE Workshop on 'What Is Green', April 20 - 21, 1994.

with other alternatives such as steel. Thorn EMI, (EMI Music, THORN, HMV, and Transaction), utilises the LCA approach (see Figure 8.1) for its reporting framework in the 1995 Group environmental report. The report layout presents the activities that have a significant impact, along with the nature of that impact, according to their respective business life-cycle stages. The reporting framework also includes targets and commitments that will respond to these life-cycle impacts. As the report audience is mainly internal, the framework of the report helps increase the employees' awareness of the life-cycle impact made by their particular part of the business. The CD plant at Uden in the Netherlands was one of the first in the world to be certified to the BS7750 standard, and other facilities are quickly following suit. The next step for Thorn EMI is to establish business-wide targets and set minimum performance standards for areas creating the greatest impact.

Sustainable Development Business Charters

As mentioned in Chapter 2, the World Commission on Environment and Development (WCED), in its publication, 'Our Common Future', emphasised the importance of building economic development strategies based on environmental conservation and sustainable development concepts. One way the business sector has responded to WCED's call is through the Paris-based International Chamber of Commerce (ICC), by setting up a task force of business representatives to create a 'Business Charter for Sustainable Development'. This Charter includes 16 principles for environmental management which guide businesses by suggesting ways to promote the principles of sustainable development through their actions.

The ICC Business Charter for Sustainable Development was launched in 1991 at the Second World Industry Conference on Environmental Management. The main objective of the ICC has been to ensure that the widest range of enterprises commit themselves to improving their environmental performance in accordance with sustainable development principles, to establishing management practices that will bring about these improvements, to measuring their progress, and to reporting this progress in a way that is appropriate both internally and externally.

The 16 principles of the ICC Business Charter for Sustainable Development are cited below. Over 1,000 international corporations have come to this consensus and agreed to make a commitment to these principles.[23] Companies should also note that the ISO 14000 requirements stipulate that all commitments to principles or charters will also be audited as part of the certification procedures.

[23] International Chamber of Commerce, ICC, "Business Charter for Sustainable Development", 1991.

Figure 8.1[24] *Business life-cycle stages for EMI Music*

[24] Excerpt from *Focus on the Environment 95*, Thorn EMI, 1995.

1. Corporate policy To recognise environmental management as being among the highest corporate priorities and as a key determining factor for achieving sustainable development; to establish policies, programs and practices for conducting operations in an environmentally-sound manner.

2. Integrated management To fully integrate these policies, programs and practices into each business as an essential element of management in all its functions.

3. Process of improvement To continue to improve corporate policies, programs and environmental performance, taking into account technical developments, scientific understanding, consumer needs and community expectations, with legal regulations as a starting point; and to apply the same environmental criteria internationally.

4. Employee education To educate, train and motivate employees to conduct their activities in an environmentally-responsible manner.

5. Prior assessment To assess the environmental impact before beginning a new activity or project, and before decommissioning a facility or leaving a site.

6. Products and services To develop and provide products or services that have no undue environmental impact and are safe in their consumption of energy and natural resources, and that can be recycled, reused, or disposed of safely.

7. Customer advice To advise and, where relevant, to educate customers, distributors and the public in the safe use, transportation, storage and disposal of products; and to apply similar considerations to the provision of services.

8. Facilities and operations To develop, design and operate facilities and conduct activities taking into consideration the efficient use of energy and materials, the sustainable use of renewable resources, the minimisation of adverse environmental impact and waste generation, and the safe and responsible disposal of residual wastes.

9. Research

To conduct or support research on the environmental impact of raw materials, products, processes, emissions and waste associated with the enterprise and on ways to minimise any adverse impact.

10. Precautionary Approach

To modify the manufacture, marketing, or use of products or services, or the conduct of activities, consistent with scientific and technical understanding, to prevent serious or irreversible environmental degradation.

11. Contractors and Suppliers

To promote the adoption of these principles by contractors acting on behalf of the enterprise, encouraging and, where appropriate, requiring improvements in their practices to make them consistent with those of the enterprise; and to encourage the wider adoption of these principles by suppliers.

12. Emergency Preparations

To develop and maintain, where significant hazards exist, emergency preparations in conjunction with the emergency services, relevant authorities and the local community, recognising potential trans-boundary impacts.

13. Transfer of Technology To contribute to the transfer of environmentally-sound technology and management methods throughout the industrial and public sectors.

14. Contributing to the Common Effort To contribute to the development of public policy and to business, governmental and intergovernmental programs and educational initiatives that will enhance environmental awareness and protection.

15. Openness to Concerns To foster openness and dialogue with employees and the public, anticipating and responding to their concerns regarding potential hazards and the impact of operations, products, waste or services, including those of trans-boundary or global significance.

16. Compliance and Reporting To measure environmental performance; to conduct regular environmental audits and assessment of compliance with company requirements, legal requirements and these principles; and periodically to provide appropriate information to the Board of Directors, shareholders, employees, the authorities and the public.

Figure 8.2 ICC business charter for sustainable development

Environmental Benchmarking Networks

Several networks have been formed so that businesses can learn by comparing what is meant by 'best environmental practice', 'environmental excellence', and the managerial tools which support best environmental practices. The network can be contacted and used as a source of information for best practices and potential benchmarking partners. Examples include:

Australian Manufacturing Council, AMC (Australia) The AMC is a council whose primary objective is to improve the international competitiveness of Australian manufacturing firms. However, the AMC is a proactive council which realises that manufacturing firms have a responsibility to take proper care of the environment if they wish to remain competitive. The AMC has produced three reports on this subject: The Environmental Challenge: Best Practice Environmental Management; Report on the Study of Industry and the Environment in Europe and North America; and The Environmental Challenge: Sustainable Businesses in the 1990s. As well, the AMC has piloted a joint legislation process between government and business where both parties play an active role in setting regulation for an industry. (Steve Prytz, Assistant Director, Australian Manufacturing Council Secretariat, t: +61.620.52.05, f: +61.3.620.52.38)

The Centre for Environmental Excellence (United Kingdom) The Centre for Environmental Excellence provides executives and managers in business, government and services with information to help them improve their environmental performance. Emphasis is placed on the bottom-line benefits gained from excellent environmental management. This information is provided through best practice case studies on exemplar companies around the world, contributions from environmental experts, and items of news and valuable advice to organisations in search of environmental excellence. (Rory L. Chase, Director of The Centre for Environmental Excellence, t & f: +44.1.234.853.605)

Global Environmental Management Initiative, GEMI (United States) GEMI is a US-based group of 27 leading companies dedicated to fostering environmental excellence by businesses worldwide. Through the collaborative efforts of its members, GEMI promotes a worldwide ethic for environmental management and sustainable development to improve the environmental performance of the business through example and leadership. GEMI has produced a variety of short publications which are intended to be used as basic guides, i.e., the Environment, Health & Safety Training Primer; Environmental Reporting in a Total Quality Framework; Benchmarking for Continuous Environmental Improvement; and Finding Cost-Effective Pollution Prevention: Incorporating Environmental Costs into Decision-Making. As well, GEMI organises numerous conferences regarding environmental management issues. This group is proactively promoting environmental leadership and transferring its knowledge and expertise in environmental management to other industry sectors and businesses. (Jane McGuire, Communications Program Manager, t: +1.202.296.74.49, f: +1.202.296.74.42)

International Benchmarking Clearinghouse, IBC, (United States) The IBC is a service of the American Productivity & Quality Centre (APQC). The APQC is dedicated to working with manufacturing, service, government, education and healthcare organisations to help improve productivity and quality management. The IBC was formed to help organisations learn from industry best practices about ways to improve quality and productivity. The IBC has become the international leader for benchmarking, and now has over 300 member organisations worldwide. They began working in the area of environmentally-related benchmarking studies in 1993. (Carla O'Dell, President, American Productivity & Quality Center, t: +1.713.681.4020 f: +1.713.681.85.78)

Nordic Business Environmental Barometer, BEB, (Norway) The BEB is a project that follows a company's adaptation towards more environmentally-friendly behaviour. The aim is to track progress at the sector, industry and country levels and to compare or benchmark these results internationally. The barometer reflects the positions and attitudes of four levels in companies: the chairman of the board, the chief executive officer, a line manager, and the environmental manager. The result is a set of benchmarks that are compiled in two-year cycles. (Bjarne E. Ytterhus, Norwegian School of Management (BI), t: +47.67.57.05.00, f: +47.67.57.08.29)

Concluding Remarks

The environmental imperative is that firms can adopt environmentally-sound practices while achieving competitive advantage at the same time. It is no longer a question of having either one or the other. Environmental standards play an important role in this process. Although often set by governmental bodies, the most enlightened ones are now being developed by those businesses which proactively define the environmental agenda. Other environmental management tools are also being developed, as well as networks to share best practices - and continually redefined environmental tools and objectives.

Chapter Nine

SHAPING THE ENVIRONMENTAL AGENDA

Achieving tomorrow's best practices requires two complementary activities: a proactive stance on defining the environmental agenda and excellent knowledge of where the various environmental actors are headed. Business managers need to know the environmental agendas of the different players that have influence on the international, regional and local decision-making processes. Moreover, they should anticipate the emerging trends related to those agendas in order to proactively meet the demands of regulators, customers and suppliers - and to outpace competitors - so that appropriate strategic actions can be planned. In this chapter, we address these issues by reviewing the five following questions:

* How will a proactive approach to the environment evolve in the future?

* What is the probable agenda of international governmental organisations?

* What are the likely environmental actions of national governments?

* How can leading-edge companies increase their understanding and respond more effectively to the concerns of environmental pressure groups?

* Who are some of the other important environmental stakeholders and what are *their* interests?

Revisiting the Proactive Approach

Creating discontinuities We refer again to our diagnostic tool for environmental benchmarking (Figure 9.1). This time, we need to focus on the arrows that link 'challenges' with 'strategic response'. When we first looked at this diagnostic tool, we were interested in the flow downward - that is, from challenges to the strategic responses. We were concerned about what we were going to *do* about the challenges. In the present context, we need to focus on the fact that the arrows go in two directions: strategic responses can cause discontinuities and create expectations.

Clearly, the reaction to an external environmental discontinuity or expectation - such as the pressure on industry to stop producing and using CFCs - did lead to a variety of company responses, including leaving the industry and developing substitutes. For ICI, finding substitutes for CFCs *created* a new set of challenges. Customers, pressure groups, regulatory agencies and international bodies responded with a sharper focus on the problem, different timetables for compliance, reductions in the acceptable levels of emissions, etc. The point is that almost any set of actions taken by firms to meet environmental challenges will create a feedback loop - which will then help to reformat those challenges.

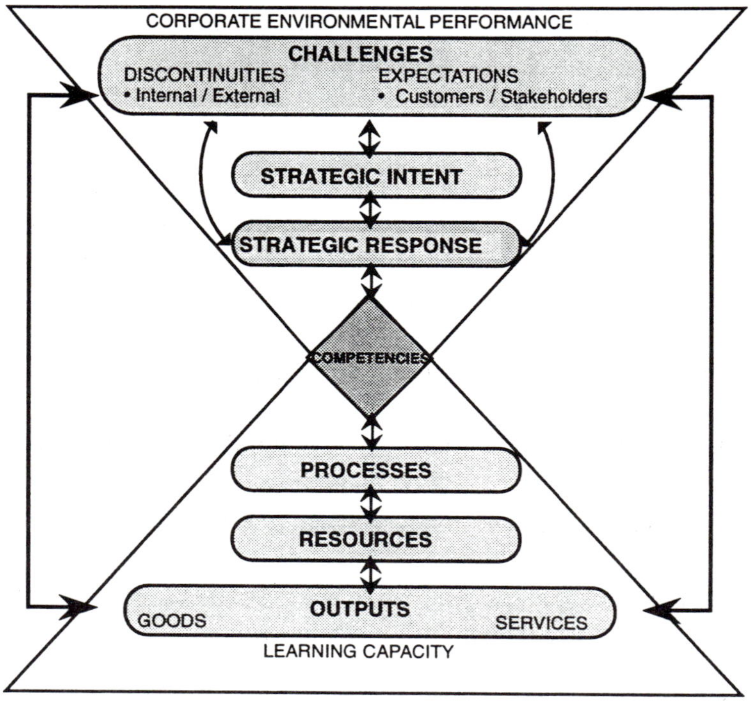

Figure 9.1. Challenges and strategic responses - the feedback loop

The resultant evolution in both environmental standards and compliance to those standards is strongly influenced by the companies that are out in front. The companies that *lead* the evolution will establish better environmental approval ratings, achieve improved ecological results, and establish the playing field for their competitors. As well, these leaders will realise meaningful economic benefits.

Environmental actors The evolutionary process is based on an interaction between a company and the various 'environmental actors' who shape the challenges. In Figure 9.1 we distinguish between external and internal discontinuities, as well as between customers, customers of the customers, and other stakeholders. A different way of thinking about the source of the challenges is to consider the present and future goals/ambitions of the following sets of environmental actors:

* *International environmental organisations which sponsor various conferences,* such as the UN. For example, at the end of the 1972 UNCHE conference in Stockholm, declarations were made, codes of conduct were formed, directions were identified for individual governmental units to follow, and specific goals were established. Some of these objectives were met early on, others later or not at all. But there *was* a feedback loop and follow-up at subsequent environmentally-related UN conferences.

* *Governmental regulators,* such as the United States Environmental Protection Agency (EPA). This group has issued many rulings over the years, such as the clean air act and

restrictions on the use of agricultural pesticides. At the time, certain companies were able to comply easily with some of these regulations, while others were not. In a few cases, the regulations were so excessively difficult to meet that they had to be revised to a less restrictive level.

* *Pressure groups,* such as Greenpeace and consumer action groups. Again, we have seen these groups change their stance on particular issues as well as on which issues are seen as being most important. Sometimes this evolution is supported by solid scientific facts, and at other times it is much more emotional. But, there *will* be evolution, and it will be strongly influenced by how businesses respond to the actions of pressure groups. As well, there will be certain firms that anticipate - indeed influence - and even *lead* this evolution.

* *Other environmental actors*, a great many forces at work to define best environmental practices. The score card is constantly changing, and those firms which adopt a proactive role will be the winners - not just in terms of best environmental practices, but also in terms of economic performance.

The key issue to be addressed is the response by companies. It is their actions which directly influence both the future requirement levels as well as the specific targets for the requirements. Examples include reductions in permissible emission levels of certain pollutants such as carbon dioxide, as well as the changed focus of these environmental actors, e.g., consumer groups' dissatisfaction with non-phosphate-based detergents concerned with reducing the amount of packaging waste being put into landfills.

If one looks closely at the evolving concerns and demands, a number of questions can be posed that are imperative for proactively developing an environmental strategy:

* Which groups - and specific actors - are the opinion leaders for various environmental issues? What is the level of knowledge about these issues in this particular company? How can it be significantly improved?

* To what kind of environmental demands is the company most and least vulnerable? Where and how should attempts to influence the evolution of environmental awareness be made? Where could a significant discontinuity be created? Could a partnership be set up to enhance a definitive environmental competitive advantage?

* Which environmental demands are most/least reasonable? How will they evolve? When there are conflicting demands, how should choices be made?

* Which challenges should be met first? What constitutes an appropriate response? When should a particular response take place? Which organisational units should tackle which opportunities? That is, how does a company decide *its* particular environmental agenda?

* How does the company adapt and ensure that best practices are multiplied?

* What should the company's priorities be? How can managers learn to resolve conflicts - particularly those between good environmental practice and market pressures?

It is clear that society is not a homogeneous entity as far as the environment is concerned. Various environmental actors have diverse value systems as well as different environmental concerns and priorities. For example, investors have become aware of the risk to the environment posed by companies which utilise toxic materials - and the potential financial liability that could occur. Thus, the financial community demands that corporate managers do whatever is necessary to avoid such risks, including using the best available technologies.

Finally, the environmental actors - and the roles they play - are changing over time. In the early years of the environmental movement, there were isolated groups and individuals who felt responsible for protecting the earth. These conservation crusaders - the first environmental stakeholders - were mainly individuals and government authorities at both national and international levels; however, as the environmental movement developed over time, other stakeholders/actors came on the scene. Moreover, the relative importance of these actors has continued to evolve. These facts need to be understood by companies trying to formulate an environmental strategy.

Environmental assessment It is not only important to know the agendas of environmental stakeholders - one also needs to understand how the various agencies make their assessments of these regulations. Figure 9.2 shows an environmental stakeholder matrix[25] where the rows refer to the criteria, metrics and assessment system used by environmental stakeholders to judge whether a company, process or product benefits the environment. It is important for managers to examine this table carefully in order to understand how environmental stakeholders think, operate and measure environmental excellence.

Environmental performance can be assessed both scientifically and qualitatively. For business and industry, it is far easier to aim for a performance that can be measured scientifically than a performance which is evaluated according to criteria that are difficult to quantify. For example, in Figure 9.2 it is possible to observe how different international organisations assess environmental performance.

The Industry and Environment Office of UNEP judges business environmental performance by looking at management from the perspective of responsibility, accountability and sustainability. The International Chamber of Commerce (ICC) and the World Business Council on Sustainable Development (WBCSD) measure environmental performance by how well companies endorse the 16 Principles contained in the ICC Charter for Sustainable Development. In both cases, the metrics used to assess the implementation of such principles are mainly qualitative. By contrast, the European Commission's criteria for environmental performance are based on how efficiently raw materials are used and how much the impact on the environment is reduced. Their metrics are mainly based on an eco-labelling scheme, with a system of assessment that measures performance against specific EC regulations. The matrix will be referred to throughout this chapter so that the reader can easily link the future agendas of various stakeholders with their current systems, and thereby assess their business and measure its environmental performance.

---

[25] Stakeholder Matrix adapted from Szekely, F., and Miller, J., "What Is Green?", *Environmental Impact Assessment Review*, No. 57, Vol. 15, No. 2/3, July 1995.

Environmental Agendas at the International Level

International organisations have had an important impact on environmental regulations and practices. That impact is growing, and companies that want to achieve environmental excellence and competitiveness need to have a clear understanding of where these efforts are headed. The importance of international organisations to the environmental debate mainly stems from three facts:

* The work of certain international organisations has been extremely prolific. They continue to develop significant environmental legislation with implications at the international, regional, and national levels.[26]

* International organisations have become the mediator and primary source of information. Organisations such as the UN are critical when it comes to resolving conflicting environmental opinions.

* International organisations have the ability to enforce judgements. After the 1992 Earth Summit, the entire international community embraced the concept of 'sustainable development' for the implementation of all future plans and programs. This has meant, amongst other things, that all multi-lateral lending institutions now require, as a condition for receiving their loans and international aid programs, that borrowers adopt and observe strict environmental and sustainable development principles.

Enforcement of existing international environmental conventions The environmental priorities of international organisations have evolved over time and will continue to do so. They have followed - and in some cases led - the concerns of society. These, in turn, are directly linked to the availability of new scientific information on the environment and the role of modern communications in spreading the message. At the present time, the international organisations are focusing on five major environmental conventions/agreements that must be met. All companies wishing to adopt a proactive environmental stance must clearly understand the present status (and level of compliance) of these conventions and agreements. Detailed knowledge here will help companies achieve best practices, in understanding what the future evolution of these initiatives might be, and in identifying concrete implications and opportunities for their firms.

[26] For example, between 1970 and 1993, the European Community developed more than 350 directives and regulations of environmental content.

Figure 9.2 Stakeholder matrix

	UNEP/ INDUSTRY & ENVIRONMENT OFFICE	EUROPEAN COMMISSION DG XI	Ethical Investment Research Service (EIRIS)	Centre Info Suisse
CRITERIA	Corporate environmental management based on responsibility, accountability, and sustainability. Voluntary endorsement of environmental conduct codes like ICC's BCSD, Responsible Care™, CERES Principles, Keidanren Global Environment Charter.	Efficient use of raw materials resources. Prevent, reduce, and eliminate pollution at source. Reduce environmental impact across the entire life cycle.	Provision of a comprehensive, public, top-level Environmental Statement dealing with all aspects of the company's environmental impact, which includes goals and implementation that can be measured objectively.	Produce products that are not hazardous from an ecological viewpoint. Implement an environmental policy and supporting information system. Conduct internal and external audits (EC guidelines). Invest in environmental protection and eco-efficiency. Signatory to CERES Principles, ICC's charter, etc.
METRICS	Use embryonic grading system to rank companies' environmental reporting practices according to 5 stages of corporate disclosure.	Give bi-annual EBEAFI awards to industry according to 5 categories: • Good Environmental Management • Clean Technology • Eco-Products • Environmental Technology Transfer • Recovery of Waste. Working on a Community-wide eco-labelling scheme.	Classify companies in 800 UK corporate groups according to 8 environmental issue areas, that can be used as exclusion criteria to generate acceptable lists of investment candidates.	Produce profiles of companies' environmental, social, and ethical performance containing both strong and weak points.
ASSESSMENT SYSTEM	Analysis of 100 pioneering companies' environmental performance reporting practices to derive 50 "key ingredients", 20 of which are core, i.e. what is feasible in 1994.	EBEAFI entries proposed by national governments are screened by national juries. Final judgments on award recipients are made by a European jury. To award eco-labels, competent bodies in Member States will assess a product's environmental performance against the principles set out in EC Regulation and criteria set for product groups.	Evaluate corporate Environmental Statements according to 4 levels: • comprehensiveness • sets and implements measurable goals • delineates responsibility for implementation • regular monitoring and review of corporate commitment to environmental management (working towards BS7750, EMAS).	Analysis based on secondary research, dialogue, questionnaires, and company visits. Use checklists to gauge performance against a range of ecological criteria. In future, may measure how far companies have implemented their signatory commitments.

	Environmental Protection Encouragement Agency (EPEA)	Council on Economic Priorities (CEP)	Investor Responsibility Research Center (IRRC)	Eco-Rating International
CRITERIA	Avoid reliance on conventional criteria like emissions and waste control. Focus on environmental compatibility of products, information policy, waste cleanup, accident and safety control, innovation, and corporate strategy.	Emissions reduction. Compliance with environmental legislation.	Eco-efficiency: creating more value with less resources and fewer environmentally-harmful outputs.	Sustainability. Use of Best Available Technology according to EC guidelines.
METRICS	Ranked the world's top 50 chemical companies against the environmental performance of their peers.	Rank the performance of companies and products as "outstanding", "moderate", "in compliance", or "poor" as compared against the average environmental performance of peers.	Environmental trend and performance indicators, including: • Hazardous Materials and Oil Spill Index™ • Emissions Efficiency Index™ • Environmental Compliance Index.™	Develop "eco-ratings" that are comparative across industry sectors, reflecting the ecological standing of a company or its products and services using a scale where "0" is environmentally-neutral; -1 to -5 is environmentally more or less harmful; and +1 to +5 achieves a net environmental gain.

	Environmental Protection Encouragement Agency (EPEA)	Council on Economic Priorities (CEP)	Investor Responsibility Research Center (IRRC)	Eco-Rating International
ASSESSMENT SYSTEM	Analysis of products, production activities, and plants. Life cycle assessment that incorporates 25 detailed criteria for environmentally-sustainable products and production according to Intelligent Products Systems.™	Use US-only data from TRI, OSHA, AIRS, and Superfund citations to profile companies' overall environmental impact, over time.	Use compliance data publicly available under US federal statutes, together with media and journal reports, to develop 1,500 company profiles.	Scientifically assess investor risk based on environmental impact from cradle-to-grave, which involves looking at a company's emissions, use of natural resources, disaster risk, logistics, infrastructure, ecological profile of products, compliance with environmental legislation, and management's capability to deal with environmentally-related risks.

	Fortune Magazine	Franklin Research & Development	WWF
CRITERIA	Amount of toxic chemical releases (adjusted for sales) and % reduction in those emissions. Participation in voluntary EPA programs (33-50). Comprehensiveness of corporate environmental program (written policy/goals; employee incentives). Legal compliance; potential Superfund liability.	Has a responsible environmental policy with targets for reducing environmental impact, and is making progress in achieving them.	Has a written environmental policy signed by top management, or "good intentions". Know the environmental impact of products and processes. Sets realistic targets to reduce environmental impact, and is making progress in achieving them.
METRICS	Published Environmental Scorecards ranking major US manufacturers from 0 (worst) to 10 (best) based on 20 differentially-weighted key performance areas.	Use a 1-5 point system where "3" represents average industry performance and a range of half or full points are assigned to develop a ranking of environmental performance reflecting how the company's performance lines up against others in its sector.	Reluctant to set down a system as still want to work with companies moving in the "right direction".
ASSESSMENT SYSTEM	Developed a "green index" based on TRI-data for the 1987-1993 period, supplemented by interviews with companies, government officials, environmental groups, and investment firms.	Building indices from toxic release data to enable comparison of emissions per unit of output, within industry sectors. Conduct general surveys followed up by discussion with companies' Environmental Managers and industry and environmental experts in the CERES coalition. Use external consultants to assess environmental impact (usually product-focussed).	No systematic approach developed. What is "green" depends on the industry. National organizations evaluate through dialogue with companies; heavy reliance on information provided by companies.

	Greenpeace	European Consumers' Organization	International Chamber of Commerce	CERES Coalition
CRITERIA	Reduction of waste. Cleaner production: • minimize energy and resource inputs • substitute materials with less environmental impact; change product design • adopt best available technologies (not EC guidelines) • make efforts to "close the loop".	Working to reduce total pollution and improve total environmental performance. Has set tough targets for source reduction, and is making progress in achieving them.	Have management practices in place to effect environmental improvement according to 16 Principles under the Business Charter for Sustainable Development (BCSD). Measure and report progress as appropriate, internally and externally.	Voluntary adoption of CERES' 10-point environmental ethic, pledging to go beyond legal compliance. Engaged in a process of continuous environmental improvement.
METRICS	No system developed but favour a quantitative approach comparative within industry sectors, evaluated on per unit of output.	No system developed.	No system developed.	Use standard, US-oriented CERES Report to document and publicly report progress in improving environmental performance.

	Greenpeace	European Consumers' Organization	International Chamber of Commerce	CERES Coalition
ASSESSMENT SYSTEM	Toxic release inventories (emissions and effluents directories) as a way to get started. LCAs, once the methodology is more rigorous. Prefer real environmental audits versus an internal review as a management tool, that looks at hazard, environmental risks, and downstream effects.	No systematic approach developed.	Evaluate performance according to: 1) Corporate priority 2) Integrated management 3) Improvement process 4) Employee education 5) Prior assessment 6) Products/services 7) Customer advice 8) Facilities/operation 9) Research 10) Precautionary approach 11) Contractors/suppliers 12) Emergency preparedness 13) Technology transfer 14) Contributing to the common effort 15) Openness to concerns 16) Compliance/reporting.	Evaluate responsible environmental stewardship based on: 1) Protection of the Biosphere 2) Sustainable Use of Natural Resources 3) Reduction/Disposal of Wastes 4) Energy Conservation 5) Risk Reduction 6) Safe Products and Services 7) Environmental Restoration 8) Informing the Public 9) Management Commitment 10) Audits/Reports.

	World Environment Center	SustainAbility	Responsible Care™	Migros
CRITERIA	Establish an overall, publicly announced environmental policy, implemented in an aggressive, sustained, innovative, and global manner. Go beyond legal compliance. Demonstrate global environmental leadership (eg. participate in sustainable development projects organized by the international community; conduct employee and public environmental awareness programs).	• creating and communicating corporate vision • developing policies and strategies • conducting environmental audits • targeting priority problems and implementing an action plan • building a common agenda with opinion formers.	Voluntary adoption of Responsible Care's™ Guiding Principles, including management standards and codes of practice, documentation, dialogue, and employee involvement. Commitment to continuous environmental performance improvement, built on existing practices and materials.	Reduction of energy, emissions, and resource consumption. Adoption of less polluting processes/products/packaging. Safe disposal.
METRICS	Annually award a Gold Medal for International Corporate Environmental Achievement.	No system developed.	Measure site performance against industry performance indicators based on 6 classes of data: 1) Environmental Protection Spending 2) Health and Safety (lost time accidents) 3) Waste/emissions ('Red List' substances discharged; disposal of 'Special' waste) 4) Distribution Accidents 5) Energy Savings 6) Complaints.	Use "eco-points" to assess the environmental burden of emissions into air and water, generation of waste, and energy consumption, comparative across products and packagings. The aim is to reduce the number of eco-points per SF1 million of sales for a defined range of goods.
ASSESSMENT SYSTEM	An independent, international jury reviews submissions made by qualifying, nominated companies, and selects award recipient.	Life cycle analysis supplemented by environmental audits and reviews.	Chief executives and site managers fill in questionnaires to evaluate performance, identify inconsistencies between policy and practice, and target areas for improvement. Evaluate progress according to achievement of 6 management practice codes.	Measure environmental protection according to 4 levels: Level 1: Economy • more economic resource deployment (energy, products, packaging) Level 2: Substitution • introduce less polluting solutions (packaging, transport, recycling) Level 3: Reduction • reduce waste, toxic waste, noise emissions Level 4: Disposal • safe disposal of non-recyclable products.

The five main international conventions are:

* The Montreal Protocol on Substances that Deplete the Ozone Layer
* The Global Climate Change Convention
* The Biodiversity Convention
* The Basel Convention on the Control of Transboundary Movements of Hazardous Wastes and their Disposal
* The London Guidelines for the Exchange of Information on Chemicals for International Trade

International organisations are increasingly demanding that companies in industrialised countries strictly observe these conventions. Moreover, the attention of these organisations is also focusing on assisting developing countries in complying with the accords. The entire UN system is strengthening the Global Environmental Facility (GEF), currently managed by the World Bank and the UNEP programs, to help developing countries change and improve their technological base in order to comply with the Montreal Protocol[27] by the year 2005. The technological changes that industry has implemented in order to comply with the Montreal Protocol - as well as its economic impact - serve as a model that will help businesses understand the environmental concerns and strategies of international organisations. When an international legal instrument is put into practice, all companies must eventually respect the resolutions in that instrument; this kind of change creates opportunities, and leading-edge companies will respond proactively. Implementation of the Protocol directly affected the interests of Du Pont and ICI, for which the production and sales of CFCs represented between 2% and 6% of worldwide revenues. These companies had two options - either to switch rapidly to the design and manufacture of alternative products or lose an important business. The latter choice had economic disadvantages as well as consequences to their reputations.

Another example of where industry had to react was in implementing the Basel Convention on Hazardous Waste Disposal. The Convention established that any international transfer of hazardous waste could take place only after the explicit approval of all parties involved. Furthermore, when the Convention became official in 1993, the twenty-three countries belonging to the Organisation of Economic Cooperation and Development (OECD) decided to ban the export of any hazardous waste to countries outside the OECD. This meant that industries generating industrial toxic waste in OECD countries where the capacity for such waste had been exceeded either had to find another OECD country willing to receive the waste or, more often, had to internalise the costs of managing toxic waste at the production site. These rules clearly changed the competitive game for many companies.

Regional governmental organisations - such as the EU, the Asian Countries region (ASEAN), the North American Free Trade Agreement (NAFTA) - also have environmental agendas which are important for business to understand and consider when making strategic decisions. For example, since municipal and solid waste is an important problem for several European countries such as Germany, France and the Netherlands, the EU has developed a number of packaging waste and recycling directives which regulate the behaviour of packaging producers

---

[27] The Montreal Protocol calls for a phasing out the production and use of five types of Chloroflurocarbons (CFCs) and three halogens. CFCs are gases which are often used in the following industries: refrigeration, construction, electronics, and packaging. Halogens are chemicals which are typically used as the basic element for fire extinguishers.

as well as packaging users. Again, a proactive environmental stance will attempt to understand the future directions of these regulations - and what problems/opportunities they represent.

The fact is that the number of international conventions and protocols are likely to increase in the future. These instruments affect both the industrialised and non-industrialised countries. In most cases, leading-edge companies are actively involved in developing the right structure to ensure implementation of the conventions within the newly-defined competitive environment. As trade agreements continue to be formulated, industries will need to meet international and regional standards of their trade partners or lose potentially large customer bases.

Promotion of international environmental management standards Standards such as the ISO 14000, EMAS, IPPC, and the Parallel Environmental Accords of NAFTA will be further promoted by international organisations in order to push business and industry into improving their environmental performance. It is in the interest of international organisations to promote these schemes so that individual countries cannot use the development and adoption of national environmental standards as an excuse to form barriers to trade.

Environmental Agendas at the National Government Level

Historically, there have been three approaches that national governments have developed to enhance environmental goals. 'Command and control' methods are where government establishes the rules on how individuals and groups must interact with the environment and then monitors actions to ensure compliance. This approach will continue to be a part of the environmental landscape, but its role has and will continue to decrease, relative to other approaches.

The second approach has been the creation of economic incentives to encourage companies (and individuals) to support environmental objectives in a cost-effective manner. One example is Denmark's tax levy on all glass wine bottles imported into the country. A required deposit on wine bottles encourages collecting and refilling them, so that less glass ends up in landfills (the import tax is refunded when the wine bottles are exported). Again, the issue for proactive business practice is to be familiar with such legislation/taxes, develop a business approach that capitalises on these challenges/opportunities, and anticipate - perhaps influence - the development of new measures - in a way that is not only good for the company, but also better for the environment.

A third avenue for national governmental action conducive to environmental conservation is to establish and promote industry sector schemes, by which industry groups determine major industry-wide environmental improvements. The added benefit is that progress can be made simultaneously for all companies in the industry as they establish a 'new playing field' within the industry. Another approach is to have closer parity among industrial sectors, with perhaps a better chance that an industry solution will be what is best for the environment. The down side is that industry groups may set the goals too low. In practice, however, this possibility has been balanced by the implied threat of command and control looming over any such actions. The achievement of sound environmental objectives by national government actions has never been easy - the proactive stance by industries and companies needs to help the governmental agencies make fewer mistakes, while promoting an even stronger improved agenda.

Policies and scientific information Governments hopefully make environmental policy decisions and establish environmental standards according to best available scientific information. Sometimes, this information is incomplete or, in the opinion of some, biased to favour a sector of society or support particular biases. As scientific evidence in the field of the environment continues to emerge and change over time, it is logical that government decisions and policies should change accordingly. For example, there is the scientific debate regarding global climate change. Many scientists do not believe that high pollution concentrations are the main cause of higher global temperatures. Others firmly believe that the concentration of carbon dioxide in the atmosphere has resulted in higher global temperatures, which in turn can be correlated with the increase in weather-related disasters - floods, forest fires, etc. These scientists are demanding stricter controls on polluters. The relevant point is that there are indeed differences of opinion when interpreting scientific evidence. Therefore, a proactive approach needs to be based on more than one scenario regarding the facts, as well as on anticipating that more than one response will be made by governmental authorities.

Conflicts of interest In attempting to forecast and understand the future direction of government decision-making concerning the environment, it is also necessary to realise the enormously different pressures that influence this arena. Some of the conflicts are between governmental units. Often such conflicts happen between international bodies and national governments that regard their own interests as being more important than those of others around the globe. This situation occurred during the 1992 Earth Summit in Rio de Janeiro; it became apparent that the interests of some governments differed from those of the international community.

Trying to determine policies about whaling provides a good example. The governments of Norway and Japan have been aggressively defending their right to continue whaling because they have developed strong economic activities in this sector. However, the international community has expressed its reservations about this activity mainly because of the danger of threatening the reproductive capacity of certain species such as the blue whale. In such situations, governments which take a decision against the general wisdom and interest of the international community are discredited and, occasionally, also suffer economic sanctions.

Conflicts of interest also exist between national, regional and local governmental units. National governments, by definition, are more concerned with and favour national interests over those of a particular community or region. When the Brazilian government decided to adopt a stringent policy to protect the rain forest, it affected the economic development of the Northeast region of the country. Traditionally, this region had been surviving through the over-exploitation of forest ecosystems. Following the government's decision, these people had to rapidly find other means of economic survival.

A government's primary responsibility is to attend to the collective demands of society as a whole. In principle, government activity in the field of the environment should respond to the needs of the people as well as those of the environment. A frequent problem is that in protecting the interest of society and the environment, government finds itself opposed by a particular individual or social group. In situations where these controversies or disputes arise, government authorities should ensure that the interests of society remain above the interests of individuals (unfortunately, this is not always the case).

There are many sources of political pressure which determine the environmental agenda of a particular government. The dynamics often result in difficult situations for industry and business. For example, in the presence of strong environmental advocates, governments tend to over-regulate industry even in situations where scientific evidence and common sense suggest otherwise. The method and rate at which recycling laws and policies have grown in Europe is a good illustration. European environmental groups have identified recycling as one of the best options for protecting the environment. Consequently, they have pushed government authorities into establishing policies and standards which require producers to utilise only materials that have a specific recycled material content. This has been done despite the fact that recycling is sometimes worse for the environment than other methods of material reuse.

Future agendas The environmental agendas of national governments tend to correspond to the degree of economic affluence held by that particular society. In industrialised countries, where some progress has been made in improving environmental performance, governments will shift their attention from the mainly regulatory approach to the promotion of more economically-efficient, market-based mechanisms. For example, they could establish and promote secondary markets for air and water pollution emissions in which economic transactions would be encouraged as a way to achieve collective environmental goals. In the EU countries, it is also expected that some form of energy tax will motivate industry to diminish energy input and decrease carbon dioxide emissions.

Governments of developing countries and Eastern Europe will have a different environmental agenda. They are still mainly concerned with promoting economic development and satisfying the population's basic needs. These governments will enforce environmental laws and standards, but at a slower rate than industrialised countries. However, over time, many developing countries will want to export their products to industrialised world markets; then, local companies will need to comply with internationally accepted environmental standards such as ISO 14000 and NAFTA.

The Agendas of Environmental Pressure Groups

The development of 'environmental pressure groups' is a recent phenomenon, the result of society's growing awareness that certain animal and plant species are in danger due to over consumption or the damage inflicted by human actions. It is useful for business firms to understand the following four related pressure group characteristics:

* Environmental pressure groups hold diverse points of view on how the environment should be conserved, based on their respective ideological backgrounds.

* Their agendas are related to the constituency they represent in their particular geo-political area.

* They tend to measure environmental performance by various, and often controversial, methods.

* More and more business firms find that their employees - and their children - are members of these pressure groups.

Diverse points of view Businesses often complain about the irrational demands that pressure groups place on them. A familiar comment is 'these people only want to put us out of business and go back to living in the stone age.' It would be a mistake to lump all environmental groups into one mindset or set of values. Environmental pressure groups are as heterogeneous and representative of numerous interests as society at large. Their strategies, membership and priorities for action change considerably. In fact, environmental pressure groups often disagree about which particular group follows the 'right' approach towards environmental protection.

To understand the nature and influence of environmental pressure groups better, it is useful to classify them according to their general interests as follows:

* *Conservation groups* The World Wildlife Fund for Nature (WWF) and the International Union for the Conservation of Nature (IUCN) are two examples.

* *Politically-active pressure groups* These include the well-known Greenpeace International, the International Committee for Saving the Whales, the Anti Nuclear Coalition, as well as many political parties such as the Green Parties in Germany and Switzerland, and the Party of the Ecologists of Canada.

* *Industry interest groups* Examples include the World Business Council for Sustainable Development (WBCSD), the Coalition of Ozone Responsible Industries, and the International Chamber of Commerce (ICC).

* *Public interest policy groups* The Center for Our Common Future, and the Club of Rome are two examples.

* *Population-oriented groups* The World Population Council, the World Health Organisation, and the World Bank.

* *Consumer organisations* Environmental pressure groups which represent the interests of consumers (such as the European Consumer's Organisation) and suppliers (the European Association of Packaging Industries) have emerged recently in industrialised countries. Their role is to educate consumers and suppliers on preferences towards products that are benign for the environment, either because of their recyclable content or because of the processes used in the stages of manufacturing and/or de-manufacturing.[28]

Environmental performance measurement systems Environmental pressure groups focus on the particular concerns of their members. Many of them have developed a system to assess whether a company's environmental efforts should be rewarded or criticised. Figure 9.2 shows a number of different systems. Politically-oriented pressure groups such as Greenpeace focus on reducing waste and having cleaner production, minimising energy and resource consumption, substituting more benign materials, designing for the environment, and adopting the best available technologies for manufacturing. In turn, organisations such as the WWF are interested in the environmental impact of products and processes, and they demand that companies set realistic targets to reduce these impacts.

[23] De-manufacturing is the concept or activity of separating a product into components at its end-of-life and then giving some use to the parts either for recycling or for secondary manufacturing.

Another approach is that of the European Consumers Organisation which is working to reduce *total* pollution and improve *total* environmental performance. This approach also requests that companies set achievable targets towards source reduction. Measuring environmental performance is a difficult task, because there is no absolute knowledge of how ecosystems work or exactly what is good for the environment. Despite these limitations, pressure groups judge the attitude of business and industry towards the environment by assessing the impact of their products and services throughout the entire life cycle. In addition to evaluating the environmental impact of processes, pressure groups also examine business practices such as: use of raw materials, ability to recycle products at end-of-life, how easily products can be de-manufactured and re-used as components of newer products, and how transporting and disposing products affect the environment.

Future agendas Pressure groups represent many different constituencies. It is difficult to estimate which environmental topics will capture their attention in the future. Some issues receive special attention for only short periods of time. For example, in 1995 the underground nuclear testing by China and the French Government's underground nuclear testing on the Mururoa Atoll in French Polynesia caused many pressure groups around the world to express strong disapproval and outrage that these governments ignored 25 years of solidarity on this issue.

Meanwhile, for businesses, three other topics seem to be systematically receiving increasing attention from most environmental pressure groups: global warming, biodiversity and deforestation. Therefore, these topics should be of special concern to most businesses as they are not a passing concern. Actions here need to be anticipated and opportunities carefully identified.

The Agendas of Other Environmental Stakeholders

There are several other environmental stakeholders that play an important role in influencing how businesses decide where to focus environmental efforts:

* The scientific community
* Consumers
* Industrial supply chains
* Other industry players and financial agents
* The media
* Religious groups

These stakeholder groups share common goals in their agendas, but move toward achieving them in very different ways. Their respective criteria, metrics and assessment systems are discussed in the following paragraphs, as well as presented graphically in Figure 9.2.

Scientific community When scientists provide hard facts about the fate of the environment, this information is often used by governments, industry and the public to develop and support their own environmental policies. Scientists have traditionally been the 'keepers of knowledge' and usually enjoy people's trust. Thus, when references to scientific facts or opinions are presented in the environmental debate, they tend to receive greater credibility and attention. However, scientists are often divided in their theories and interpretation of data, and its relevance to the environment. Since the beginning of the environmental movement in the late 1960s, several scientific works have had tremendous impact on people's perception of the environment. These works include:

* *Dr. Barry Commoner's 'The Closing Circle'*[29] Dr. Commoner's message is that today's environmental problems are directly linked to the technology we have created. He uses agricultural technology based on the intensive use of nitrogen and phosphorous fertilisers as an example. He claims that modern technologies are energy intensive and have broken nature's circle by altering the functioning of hydrological and other cycles. The solution proposed by Commoner is to close the circle by designing technologies which are compatible with the functioning of natural cycles.

* *Dr. Paul Erlich's 'The Population Bomb'*[30] The main argument is that the present environmental problems are due to the population explosion. Population figures are growing exponentially while the access to resources is growing arithmetically. Erlich's solution is to institute strict (and compulsory) population control programs, mainly in the poorest third-world countries.

* *NASA's scientific data on the thinning of the earth's ozone layer* After years of debate about whether the use of chemicals - such as CFCs and certain halogens - is responsible for the depletion of the ozone layer, the scientific data gathered by NASA satellites has provided the necessary evidence to link the production and use of these chemicals with deterioration of one of the largest natural systems.

The *scientific community* will continue to validate hypotheses and opinions on the environment. New evidence regarding root causes of environmental degradation continue to emerge. One example is the case of Dr. Rowland from the University of California, who warned the international environmental community over 20 years ago that CFCs were responsible for thinning the ozone layer. In August 1995, Dr. Rowland announced that the accepted scientific causes of air pollution problems in mega-cities - such as Mexico City, São Paulo and others - is incorrect. Conventional theory is that automobiles and old transportation technologies are the main sources of air pollution in mega-cities. Dr. Rowland discovered that the continuous escape of Liquefied Petroleum Gas (LPG) - from old stoves and poorly built and maintained domestic gas containers - was, in fact, lowering the ozone in those cities. Industry needs to be aware of such findings that lead to a better understanding of why specific environmentally-related phenomena occur. New findings need to be integrated into business strategies in a timely manner if a competitive advantage is to be gained from this new knowledge.

Consumers The consumer movemen*t* is one of the most significant forces for environmental improvements. The consumer movement existed before the environment became a key concern to modern society. It was an American lawyer named Ralph Nader who really launched the movement by advocating the public's 'right to know' about the safety and/or potential toxicity of products being purchased. Nader revealed that some companies were sacrificing the consumer's safety for economic gains.

---

29 Commoner, B., *The Closing Circle*, New York: Alfred P. Knopf, 1971.

25 Erlich, P., *The Population Bomb*, Stanford, CA: Stanford University Press, 1970.

When certain consumer organisations started working for stricter rules to protect the well-being of workers, some industries saw a great potential for developing new businesses and products which could satisfy environmental demands. They began competing for consumers by offering 'green' products and services that not only would satisfy consumer demands but even anticipate them. They were taking the concept of being customer oriented a step further: Pleasing customers in an environmental context is the ultimate goal.

It is well known that rising standards of living have often led to increasing public concern about environmental issues. When *Which?*[31] surveyed its subscribers in late 1986, nine out of ten reported that they were worried about pollution and other environmental problems. *Which?* decided to inform its customers that 'cleaning up pollution isn't cheap' as it means more expensive electricity, cars, water or higher taxes. A good number of those consumers, after assimilating the message, started showing a preference for buying products and services that minimise any negative impact on the environment.

As shown in Figure 9.2, consumer organisations such as the European Consumer's Organisation (ECO) influence the business sector by applying its measurement system to assess good or best environmental practices. The ECO's criteria, metrics and assessment systems are in relative terms. Although the organisation's final objective is to inform consumers about the environmental impact associated with their commercial choices, the criteria for measuring good environmental performance consist mainly of assessing the efforts that producer/service providers make towards reducing pollution. Although no systematic approach has been developed for a truly effective environmental assessment system, environmental measurement systems do exist, and one can expect to see steady improvement in the near future. On the other hand, consumer organisations wishing to elicit proper responses from the business sector need to be as specific as possible in the targets and tools they utilise.

The basic message for company managers relating to consumers as environmental stakeholders is that people are increasingly demanding good environmental performance from products and services. As consumers become more environmentally informed, their concerns have moved closer to life-cycle thinking. However, at this time only a small percentage of the population is prepared to pay more for better environmental performance. The message for business is that products and services need to be competitive in price, while simultaneously reducing relative environmental impact. A valid question regarding this statement is often asked: Is this possible? Is it possible to treat environmental performance as an internal business cost such as quality or the cost of raw material? The answer is yes - this is the challenge that leading-edge firms are meeting. It is not only good for the environment, it is good business.

Consumers will increasingly demand more detailed information about the environmental impact associated with the services and products they purchase and use. Moreover, consumers will continually increase their environmental demands. Consumers will also expect industry to create environmentally-friendlier products and services for the same price as those that are not so green. Satisfying these demands will separate the winners from the losers.

---

[31] A British Magazine

Industrial supply chains Companies have also become an important environmental stakeholder by encouraging their *contractors and suppliers* to have practices, products and materials that are consistent with the enterprise's environmental principles. A good example can be seen in the automotive industry. Nissan Motor Company Ltd. was one of the first to establish a recycling program which was run by the Recycle Promotion Committee. This committee met with Nissan suppliers to explain Nissan's policies and secure their cooperation in marketing recyclable plastics.

Canadian Pacific Hotels & Resorts[32] is another example of a large corporation which has exerted pressure on suppliers and contractors. In its 'Green Partnership Guide', recently sent to all Canadian Pacific Hotels, the company's Environmental Affairs Department encourages its hotels in no uncertain terms to: '...be sceptical of 'green' labels and manufacturers' designations'. The company asks managers to go through the following exercise in order to make sure that the supplies for its hotels have an environmental content.

1. Begin with a few fundamental questions about every purchase:
    - Do we really need this? If the answer is NO, don't buy it.
    - Does this product damage the environment? If the answer is YES, don't buy it.
    - Is there a more nature-friendly alternative? If the answer is YES, buy *that* one.

2. Buy in bulk to reduce costs and packaging waste. Don't buy individual or single-serving packages, particularly for restaurant and kitchen supplies.

3. Reject packages and containers that cannot be returned, re-used, or recycled.

4. Don't buy single-use, throw-away or 'disposable' anything, whether it's for a newsstand, a gift shop, or for hotel supplies.

5. Always opt for products that are natural, organic, chemical-free and synthetic-free.

6. Buy recycled items whenever and wherever available.

A similar example is found at S.C. Johnson & Son, Inc. The Wisconsin-based cleaning products manufacturer regards working with external business partners, particularly suppliers, as being essential to advancing environmental progress. In 1991, the president of S.C. Johnson hosted an environmental symposium entitled 'Partners Working for a Better World', which was attended by 57 of the company's 70 supplier organisations. The conference included presentations by environmental celebrities from government, non-governmental organisations and pressure groups. Following the presentations, the participants broke into groups to identify broad steps for action that would support mutually-stated environmental goals. The discussion focused on bio-degradation, source reduction, recycling and volatile organic compounds.

Xerox is another company example, as it has made a special effort to develop comprehensive multinational supplier packaging programs to standardise packaging and pallets, and provide

---

32 Adapted from Willums, J.O., and Golüke, U., *The Greening of Enterprise 1992*, International Environmental Bureau of the ICC, 1992.

for their re-use. This program expects to avoid about 10,000 tons of waste and save up to $15 million annually.

Numerous other companies are working with suppliers, as they increasingly realise that it is the entire product/service life cycle that needs to be addressed. The same philosophy is presented in the MIBE environmental wheel (Figure 3.4 in Chapter 3). These examples prove that many stakeholder benchmarks regarding supply chain management already exist and can be improved upon to achieve a competitive advantage.

Industry is moving from a mainly compliance type of environmental response to a competitive approach. More and more industries will identify the environment as a tool for becoming more cost effective, for conserving the environment and sustaining business over the long term.

Other industry players and financial agents Environmental regulations and environmental demands are a source of enormous profit for some industry sectors - such as the environmental technology and engineering industry. This industry includes everything that relates to pollution control technology, the operation and maintenance of environmental capital goods (equipment, treatment plants), and environmental services such as waste collection, recycling and environmental consulting. It is in the interests of these types of companies to see stricter environmental laws, policies and regulations. Thus, it is common to see companies in this sector lobbying for the adoption of more stringent environmental legislation and standards.

The packaging industry offers a good example of this situation. The glass industry actively lobbies for government regulation on the use and disposal of plastics, because stricter environmental legislation on plastics enhances the glass industry's chances of being chosen as a cheaper, more environmentally-friendly substitute. Thus, other industries should be considered as an important environmental stakeholder - not necessarily because their main concern is the environment, but because the success of their business directly depends on having higher and tighter environmental restrictions for others.

Financial institutions are not only beginning to offer 'green investment' portfolios, but they are also including environmental criteria in loan agreements and investment decisions. At National Westminster Bank in the United Kingdom, business customers' environmental records are examined when making loan decisions. Companies are assessed, based on a set of criteria which includes the quality of their corporate management. The bank also looks for evidence of a long-term view with regard to environmental performance, and companies are automatically graded into low, medium, or high environmental risk loan categories based on their industry sector. This assessment is more than an interest in better environmental practices. It also is a matter of financial risk. Companies with good environmental practices will ultimately be better credit risks.

For environmentally-related business sectors, the future looks promising as certification to national and international environmental standards becomes a minimum requirement for doing business. Competitive 'greener' alternatives will continue to be introduced as companies begin to feel more pressure from environmental stakeholders. Financial institutions will continue to put pressure on businesses to decrease their exposure to future environmental liabilities. They will play a crucial role in bringing the environment closer to the strategic planning phase of corporations and make it an additional requirement in financial reporting structures.

Media The media is an important environmental stakeholder that sometimes informs the public in a self-serving manner. The media's power lies in the fact that environmental news can be delivered in a such an emotional manner that scientific fact and technology no longer matter. Indeed, the concept of a big company doing harm to the environment or to people who support the environment has considerable emotional appeal. The media influences the environmental opinions that people form. For good business practice, this indicates that environmental news - good and bad - needs to largely come from the companies who create the news. Any attempt to hide the truth will usually lead to bigger problems in the long run.

When Exxon caused an oil spill in the Gulf of Alaska in 1989, the media consistently featured the graveness of this situation. Why did the Exxon incident become an ongoing media event, when it was not the worst oil-related disaster in history? The answer could lie in Exxon's poor response to this incident. From the first news of the oil spill, Exxon was extremely reluctant to accept any wrongdoing. This reaction infuriated many people among the press as well as the public. Consequently, some parts of the media perpetrated an undeclared war on this company for not behaving responsibly towards the environment and the public.

Another example was the confrontation between Shell Oil International and Greenpeace - over the company's plan to dump the Brent Spar oil collection platform at sea. In this case, Greenpeace was the media's main source of information. The media consistently portrayed the company as insensitive to environmental issues and not telling the truth about the environmental consequences of its plans. The Greenpeace campaign was successful. It was able to prevent the use of the ocean as a landfill site - even though it was later proven that Shell Oil, not Greenpeace, did have the right information about the hazardous substances contained within the Brent Spar.

The media can be expected to continue using the environment as a 'hot topic'. However, as scientific facts become more readily available, and businesses more openly disclose environmental performance, more reasoned reporting should ensue. The media is and should be part of a policing group which communicates environmental issues and identifies wrongdoing.

Religious groups Various religious groups have been extremely effective in protecting the environment, participating actively in both international and local debates. Christians from very diverse backgrounds - from Roman Catholics to the Franciscans to the World Council of Churches - tend to be concerned about ecology. They look for solutions to the planet's current-day problems by seeking the root cause in a biblical context. The Dalai Lama, speaking for the Buddhist religion, has frequently explicitly stated the Buddhists' mission to protect the environment. The Jewish faith has always embraced protection of the environment. Most of the Jewish ceremonies, laws and resources related to creation consider the environment with great respect. The guiding principles regarding planting and harvesting are strictly spelled out: people should do what is necessary to maintain what exists, but the land must not be exploited. Similarly, the laws that guide Muslim life reflect an inherent concern for ecology. Muslims have traditionally had a broad perspective on issues as wide-ranging as animal husbandry, desert reclamation, international trade and science. Many verses of the Quran focus explicitly on different aspects of the environment - particularly on the importance of water.

It is important for a company to understand various religious agendas, particularly when building operations in countries where religious beliefs have been integrated into governmental systems. Religious organisations can be expected to focus on the environment for unifying their communities. In some cases, environmental issues will encourage more traditional values.

Concluding Remarks

The environmental agendas of some actors are readily accommodated, while those of others occasionally can play havoc with an organisation's products and processes. It is in a company's best interest to be aware of those agendas that have the highest probability of having an impact on its business interest. The most critical groups could be local or global, but it is certain that companies which focus on only one group's concerns are vulnerable to being caught by another one. It is important that all employees take an active role in understanding who the most important environmental stakeholders are for their individual responsibilities. This will enable a company to 'feel the pulse' of the concerns of many environmental actors. That is, scanning the environment for potential problems/opportunities needs to be everyone's job. Open dialogue with these actors can reinforce future environmental strategies and actions - but companies must never expect a relationship without confrontation. Best practise includes building a relationship of respect for other concerns, be they financial or ecologically based, as well as to respect the role of such groups in encouraging better environmental practices.

Chapter Ten

SUMMARY AND CONCLUSION

In this last chapter, we have summarised the most critical points for business managers to consider. We have reviewed each chapter, highlighted the primary messages for business managers, derived the key learning points, and formulated recommendations for future actions.

Chapter 1 - The Environmental Imperative

Messages for business managers Many business managers believe that achieving excellent economic results and protecting the environment are conflicting concepts. However, evidence now exists that fulfilling these demands simultaneously is not only possible but can also be an important means of gaining competitive advantage.

Firms that carry out environmentally-sound business development in a cost-effective manner will dominate their industries, while those which treat environmental issues as being secondary or only a form of good citizenship will lose out.

Learning points Benchmarking is a management tool that can be explicitly applied to improving environmental performance in ways that will help companies increase their competitiveness.

Benchmarking in an environmental context is not a static activity: it is a process of continuous improvement.

Recommendations Management teams should enhance the environmental imperative: *We will leverage better environmental performance to achieve a competitive advantage.*

Chapter 2 - The World Responds to the Environmental Movement

Messages for business managers Four primary approaches have been used to achieve better performance: command and control, market-based incentives, industry self-regulation, and a visionary response for maintaining sustainable development. Companies which have pursued a visionary approach towards the environment, and have thus contributed to setting the environmental agenda in their particular sector, have become the industry leaders.

The term 'sustainable development', as applied to the environment and business, addresses the critical interrelationship between economic growth, environmental quality and social welfare. Industries that embrace the concept of sustainable development are using a 'visionary' response towards society's environmental demands.

Fulfilling society's environmental demands does not imply that business activity will decrease. In fact, these demands can be turned into forces which will provide many business opportunities. Adopting a 'proactive' approach towards the environment means that two goals can be attained simultaneously: doing better business and conserving the environment.

Learning points The command and control model for environmental management is well known and understood by most stakeholders. However, international standards tend to be difficult to design and enforce. Command and control is often applied when it is too late, as illustrated by the case of stringent air pollution standards for the automobile industry in areas such as Mexico City.

Market-based incentives can be a useful tool for environmental management. However, since the marketplace is not yet sufficiently developed to benefit from this approach, it might function inefficiently and ineffectively.

Industry-self regulation is the best way to improve environmental quality; however, the complexity of particular environmental problems may exceed the capacity of a single industry sector to solve an environmental problem thoroughly. Proactive environmental self-regulation has proven useful in some industry sectors. For example, industry groups such as the chemical and oil industries have recognised that it is definitely in their interest to determine their own environmental standards - based on a self-assessment of what is most economically viable - and to proactively reach environmental goals which exceed those established by regulatory agencies.

Recommendations Business managers should shorten the 'promenade' which will bring their company from taking a defensive position towards the environment to pursuing a visionary one. In this way, they can help their company become the leader in its industry sector.

Business managers who are looking for ways and means to increase their company's competitive advantage should look at the environment as a realistic source for developing and achieving this goal.

Chapter 3 - Benchmarking Environmental Performance

Messages for business managers The objective of environmental benchmarking is to identify and sharpen the abilities and attitudes that a company needs in order to excel in business and environmental performance simultaneously. The benchmarking activity should focus on identifying companies that have increased their business competitiveness by improving environmental performance.

The key issue in environmental benchmarking is that it must develop the fundamental infrastructure to support environmental goals. The objective is to learn how to design and operate this infrastructure in a much better way. Environmental benchmarking must focus on processes and the practice of processes.

The scope of environmental benchmarking needs to encompass all areas of the firm's activities. Improvements are not limited to manufacturing processes; breakthrough performance is often associated with taking a much wider perspective on environmental issues.

Learning points The benchmarking activity must emphasise the practice of processes and not focus only on measurement. Environmental issues need to be explicitly coupled with existing improvement efforts. Total Quality Management (TQM) is an excellent example.

An valuable extension of the TQM philosophy applies Deming's 14 principles of quality in an environmental context. Those who have already adopted the Deming philosophy have quickly realised its value.

Recommendations We recommend that business managers build on their experience in existing improvement programs, such as TQM, and apply them to environmental management improvements.

Managers should focus their efforts on new processes that will achieve environmental objectives and on the learning gained from benchmarking best practices, as they provide opportunities to excel in performance.

Chapter 4 - Phase 1: Discovery What are the Company's Environmental Challenges?

Messages for business managers To conduct environmental benchmarking, the business manager must understand, in concrete terms, the environmental challenges facing the company and the detailed responses that must be made to meet those challenges.

To develop an effective company environmental action plan, the people who will implement the change need to understand how the environment is linked to traditional business goals and how it can help the business become more efficient and competitive.

Learning points A manager can codify and prioritise the key environmental challenges facing the company by using the Environmental Business Transformation Diagnostic. This tool focuses on assessing the current set of internal and external challenges for improving the organisation's environmental performance.

The belief that protecting the environment is an expense rather than an investment with a payback has inhibited many business managers from making significant changes. For environmental benchmarking to be beneficial, it is vital to identify competencies that clearly correspond with the organisation's environmental business strategy.

Using a cross-functional management team of eight to ten people has proved to be an effective way to launch the 'discovery' phase of the environmental benchmarking process.

Recommendations When making an environmental diagnosis, it is of paramount importance that the people participating in the final benchmarking study also be involved in developing the competency map.

Chapter 5 - Phase 2: Assessment
How to Select Processes and Environmental Benchmarking Partners?

Messages for business managers Many of the company's competencies and their underlying processes need to be re-engineered in order to obtain an environmental competitive advantage.

Benchmarking with companies in other industries - especially their environmentally-related processes - is extremely useful, as this is where the 'breakout' ideas are generated. However, a company can also benchmark its processes and practices internally.

There are four important attributes of excellent environmental performance which a company manager must pursue: attitude, awareness, accountability, and action.

Benchmarking is an art that needs to be practiced intelligently in order to achieve best results.

Learning points The appropriate selection of the benchmarking team is a critical element in the potential success of the environmental benchmarking process.

A decade of benchmarking allowed the Japanese to adopt excellent business practices which helped them improve their operations and eventually be recognised for designing superior - more durable, functional and reliable - products.

There are several common mistakes that can prevent the benchmarking team from accomplishing more than only 'recreational benchmarking'. For example: 'feel good' trips, vague questions/goals, too broad a scope, focusing on metrics not processes, no team commitment, neglecting the homework, poor selection of the benchmarking processes and industries, and inadequately following up the activities.

Recommendations Business managers should adopt excellent environmental practice (EEP) as a goal.

The company/ies selected for benchmarking must have an organisational fit in order to ensure higher relevance and successful adoption of the benchmarking lessons.

The team should invest maximum 'homework' time and exchange information prior to the best practice visit.

Chapter 6 - Phase 3: Action
Conducting Visits and Developing Implementation Plans

Messages for business managers Discovering the critical gaps while doing environmental benchmarking is one of the most important results of the process. The team must determine the gaps that will lead to improvement, that will set goals and action plans.

Business managers have to address three main questions if they intend to conduct appropriate environmental benchmarking: 'what' to benchmark, 'why' benchmark a process, and 'how' certain processes work in a company.

Another key goal of excellent benchmarking is looking for ways to learn when others visit - as well as gaining something from visiting elsewhere. Slightly different skills are required, but clearly there is *always* an opportunity for complementary learning.

Learning points The most productive way to conduct environmental benchmarking visits is by assigning specific tasks to every member of the benchmarking team. Each person is responsible for driving the analysis to a particular block of questions. It is good practice to have more than one team member taking notes simultaneously, to ensure that all information is recorded.

Discovering the critical gaps is one of the most important results of the benchmarking process, as these gaps lead to setting goals and action plans.

The goals and action programs formulated by the benchmarking team need to be ambitious - a 50% reduction is a reasonable goal. But, these recommendations should never be interpreted by the present process owners as an indictment of their practices. As well, they may need to focus more on 'how' rather than on 'what'.

Timing is important, as the level of commitment to making the changes is highest just after the benchmarking site visit. Including more of the process owners may initially slow down the planning, but the results are usually much easier to implement and more acceptable to the end users.

Recommendations The results - including a description of the benchmarking journey - need to be communicated widely so that commitment will be extensive, thereby enhancing the chances that implementation will be quickly achieved and the next steps can be formulated.

Choosing the implementation team members is critical: for implementation to proceed, the people on the team must be good change agents and be respected by their peers as well as by upper management. Successful wins by early adopters will give positive endorsement to the new environmental imperative.

Chapter 7 - Phase 4: Continuous Improvement

Messages for business managers Benchmarking in its widest sense is concerned with learning from others' experiences. Many types of learning can also be shared. One example is 'practice sharing' the company's more general issues/concerns. Managers should identify the environmental practices that their business units perform with excellence in order to benefit from these internal experiences. Managers need to overcome the competitive atmosphere which traditionally exists among business units.

Practice sharing must be treated as a two-way learning experience rather than just as giving away good business ideas.

Benchmarking needs to be seen as part of the methodology for achieving continuous improvement, as being an important contributor to the long-term success of the company.

Learning points It is good practice to be ahead of environmental challenges. For example, when the U.S. Environmental Protection Agency first issued its regulations for auto emissions, American firms hired lawyers to overturn the regulations - while Japanese firms hired engineers to *meet* the regulations.

Managers who closely follow the evolution of the environmental movement are able to anticipate society's environmental demands in different socio-political settings. This knowledge can be converted into actions which will benefit the company's business operations and prevent unwanted environmental risks.

Recommendations Instead of just responding to the environmental challenges, gaining true competitive advantage will be by finding ways to anticipate valuable environmental needs in various fields, then setting the agenda or creating environmental discontinuities, and doing so more rapidly than competitors.

Chapter 8 - Environmental Standards and Management Tools

Messages for business managers In the process of improving the environmental performance of products and processes, some companies have gone beyond becoming more competitive; they have actually set the environmental standard for the entire industry. Leading-edge companies not only *comply* with environmental standards - their actions help *establish* the standards.

Several complementary environmental models and concepts are available for use in conjunction with environmental benchmarking, e.g., regional environmental management standards, national and international environmental standards and schemes, life-cycle assessment tools, sustainable development charters, and environmental benchmarking networks.

There is a strong correlation between international and regional environmental standards and international and regional trade. Business managers of international companies must know, in detail, the environmental regulations of each socio-political area where their companies produce, serve or sell products and services.

Learning points Although most international and regional trade agreements, such as the EU and NAFTA, have explicitly stated that environmental standards should not constitute barriers to trade, international and regional environmental standards are already having an important effect on international trade. Companies which want to participate in specific regional markets must learn to comply with the environmental rules and regulations of these markets.

The existence of national voluntary environmental management standards such as the British BS 7750 has fostered competitiveness among peers in industry sectors and has enabled companies to inform the public about their commitment to the environment. Companies that are complying with such standards have proof that good standard procedures for dealing with the environment are in place.

ISO 14000 is an international voluntary environmental management standard that verifies when a company is using an environmental management system which prevents or detects environmental problems, regardless of where the company operations are located. The standard focuses more on management practices than on compliance. The standard integrates environmental responsibility into everyday corporate decision-making. The existence and adoption of ISO 14000 is increasing in importance for companies with international operations.

Recommendations Business managers should adopt, as soon as possible, ISO 14000 certification standards, as well as the International Chamber of Commerce Charter for Sustainable Development.

Business managers must develop a better understanding of 'best environmental practice', 'environmental excellence', and the managerial tools which support best environmental practices.

Business managers should make more and better use of existing environmental benchmarking networks.

Chapter 9 - Shaping the Environmental Agenda

Messages for business managers Today's best environmental management practices are not necessarily tomorrow's best practices. The future requires two complementary activities: a proactive stance on defining the environmental agenda and excellent knowledge of where the various environmental actors are headed.

Business managers must anticipate emerging environmental trends related to particular environmental agendas in order to proactively meet the demands of regulators, customers and suppliers, as well as to outpace competitors - so that the appropriate strategic actions can be planned.

There are many different 'environmental actors', all pursuing environmental agendas which reflect the values of their constituencies. Examples of these actors include international environmental organisations, government regulators, pressure groups, and other actors such as financial agents and academic institutions.

Learning points These various actors have different criteria, metrics and assessment systems to determine what is 'good' for the environment.

Environmental pressure groups hold diverse points of view on how the environment should be conserved, based on their respective ideological backgrounds.

Most business managers will find that they have some close relatives who support or participate in different environmental pressure group activities.

Recommendations Business managers must learn to identify which groups - and specific actors - are the opinion leaders for the various environmental issues in their industry sector.

Business managers should know which environmental demands are most and least likely to affect their company.

Business managers should be aware of the different criteria, metrics and environmental assessment systems used by the most relevant environmental actors; they need to work with them to improve these criteria and assessment systems, as well as to promote consistent use of the evaluation tools.

Closing Remarks

Modern society demands a great deal from its business leaders. These demands often seem irreconcilable, including the need to maintain or achieve excellent economic results while at the same time taking a major role in protecting the environment. Increasingly, however, achieving these apparently conflicting demands is proving to be not only be possible - it is being used to obtain a critical source of competitive advantage.

The primary objective of this book has been to provide a path whereby managers can incorporate environmental dimensions into their daily business practices in ways that will allow them to conserve the environment *and* increase their long-term competitive ability. The main tool we propose is a benchmarking process that begins with the environmental business transformation diagnostic.

This diagnostic and process has proven to be very useful, enabling companies to compare their environmental practices with those of others which have achieved excellent environmental performance. Learning to leverage solutions to environmental problems is one way to gain competitive advantage.

In the end, it is the environmental imperative that is all-important. The fundamental message of this book is that it is possible to carry out environmentally-sound business development which is cost effective and competitive. Those firms which do so will dominate their industries, while those which treat environmental issues as being secondary or as only a form of good citizenship will lose out.

USEFUL READING LIST

1. Australian Manufacturing Council (1992)
   The Environmental Challenge: Best Practice Environmental Management, Melbourne, Vic: AMC.

2. Camp, R. C. (1989)
   Benchmarking: The Search for Industry Best Practices That Lead to Superior Performance, Milwaukee, WI: ASQC Quality Press.

3. Stephan Schmidheiny with The Business Council for Sustainable Development (1992)
   Changing Course: A Global Business Perspective on Development and the Environment, Cambridge, MA: The MIT Press.

4. Fuller, Karl; Harrison, Mike, James, Peter; Whelan, Bob (1994)
   "Theme: Benchmarking", *Environmental Excellence*, IFS International Limited, pp. 22-27, December 1994.

5. GEMI (1993)
   Total Quality Environmental Management the Primer, Global Environmental Management Initiative, Washington DC.

6. GEMI (1994)
   Benchmarking: The Primer Global Environmental Management Initiative, Washington, DC.

7. GEMI Conference Committee, Conference Proceedings, March 24 - 25, 1993
   Corporate Quality Environmental Management III: Leadership - Vision to Reality, Global Environmental Management Initiative.

8. Hammer, Michael; Champy, James (1993)
   Re-engineering the Corporation: A Manifesto for Business Revolution, New York, NY: Harper Collins.

9. Porter, M.E., van der Linde, C. (1995)
   "Green and Competitive", *Harvard Business Review*, pp. 120-134, September-October 1995. and Letters to the Editor (1995) "Green and Competitive", Harvard Business Review, pp. 194-208, November-December 1995.

10. Szekely, F., Miller, J. (1995)
    What Is Green?, Environmental Impact Assessment Review, No. 57, Vol. 15, No. 2/3, July 1995.

11. SustainAbility, UNEP (1994)
    Company Environmental Reporting: A Measure of the Progress of Business & Industry Towards Sustainable Development, UNEP Technical Report No. 24, ISBN 92-807-1413-9.

12. Vollmann, T.E. (1996)
    The Transformation Imperative: Achieving Market Dominance Through Radical Change, Boston, MA: Harvard Business School Press.

APPENDIX A

50 Reporting Ingredients According to UNEP

THE 50 REPORTING INGREDIENTS

I. Management & System

1. CEO Statement
2. Environmental Policy
3. Environmental Management System
4. Management Responsibility
5. Environmental Auditing
6. Goals and Targets
7. Legal Compliance
8. Research and Development
9. Programmes and Initiatives
10. Awards
11. Verification
12. Reporting Policy
13. Corporate Context

II. Input/Output Inventory

Inputs

14. Material Use
15. Energy Consumption
16. Water consumption

Process Management

17. Health & Safety
18. Environmental Impact Assessment & Risk Management
19. Accidents & Emergency Response
20. Land Contamination & Remediation
21. Habitats

Outputs

22. Wastes
23. Air Emissions
24. Water Effluents
25. Noise and Odours
26. Transportation

Products

27. Life Cycle Design
28. Packaging
29. Product Impacts
30. Product Stewardship

III. Finance

31. Environmental Spending
32. Liabilities
33. Economic Instruments
34. Environmental Cost Accounting
35. Benefits and Opportunities
36. Charitable Contributions

IV. Stakeholder Relations

37. Employees
38. Legislators & Regulators
39. Local communities
40. Investors
41. Suppliers
42. Consumers
43. Industry Associations
44. Environment Groups
45. Science & Education
46. Media

V. Sustainable Development

47. Global Environment
48. Global Development
49. Technology Cooperation
50. Global Standards

APPENDIX B

Documents Supporting Bristol-Myers Squibb's Environmental Mission

The Bristol-Myers Squibb Pledge

To those who use our products...

We affirm Bristol-Myers Squibb's commitment to the highest standards of excellence, safety, and reliability in everything we make. We pledge to offer products of the highest quality and to work diligently to keep improving them.

To our employees and those who may join us ...

We pledge personal respect, fair compensation, and equal treatment. We acknowledge our obligation to provide able and humane leadership throughout the organization within a clean and safe working environment. To all who qualify for advancement, we will make every effort to provide opportunity.

To our suppliers and customers...

We pledge an open door; courteous, efficient, and ethical dealing; and appreciation of their right to a fair profit.

To our shareholders...

We pledge a company-wide dedication to continued profitable growth, sustained by strong finances, a high level of research and development, and facilities second to none.

To the communities where we have plants and offices...

We pledge conscientious citizenship, a helping hand for worthwhile causes, and constructive action in support of civic and environmental progress.

To the countries where we do business...

We pledge ourselves to be a good citizen and to show full consideration for the rights of others while reserving the right to stand up for our own.

Above all, to the world we live in...

We pledge Bristol-Myers Squibb to policies and practices which fully embody the responsibility, integrity, and decency required of free enterprise if it is to merit and maintain the confidence of our society.

Bristol-Myers Squibb

Corporate Purchasing Guidelines
(Adopted 8/8/91)

It is company policy to protect the health, safety and quality of life of its employees and the public, and to exercise responsible stewardship of natural resources that may be impacted by Company activities.

Without compromising the quality and competitiveness of our products or workplace, Purchasing is to share in the responsibility for minimizing any adverse environmental impact directly associated with materials and services purchased by the Company.

At a minimum, no purchasing decisions should be made before considering the following:

Source Reduction: Can less material be used through value analysis?

Recycled Availability: Is the item available in recycled form?

Recyclability/Reusability: Is the item recyclable or reusable, or can it become so given sufficient demand?

Renewable Resources: Can the material be replaced by an environmentally preferable renewable resource?

Hazard Reduction: Can a less hazardous material be substituted?

Process Waste Reduction: Can in-process waste be generated by the material or its use be minimized?

Handling and Disposal: Does the item, its intermediaries or wastes require special handling and/or disposal?

Environmental Technology: Does the item incorporate environmentally preferable technology?

Energy Efficiency: Have products, processes and waste disposal associated with the item been evaluated relative to environmental and financial energy costs?

Suppliers: Does the supplier share the Company's environmental goals and objectives, or can the Company's purchasing requirements be used to encourage a supplier to improve its environmental sensitivity?

Towards the successful implementation of these guidelines, Purchasing is expected to:

- Stay abreast of relevant product-related environmental information, innovations, and trends.

- Work closely with Company clients to identify environmentally preferable purchasing options.

- Notify and regularly update suppliers as to the Company's environmental commitment and objectives.

Any questions or requests for information regarding these Guidelines should be directed to your local Purchasing organization.

Bristol-Myers Squibb
Corporate Packaging Committee

Environmental Guidelines for Package Development
(4/3/90)

Environmental awareness is the responsibility of all Bristol-Myers Squibb employees, including those who participate in the design, development and approval of packaging materials and product packages. We must work towards reducing the impact of packaging on solid waste without compromising package effectiveness for the consumer. The following items are to be addressed during each package development and reviewed for each current package.

Source Reduction

Consideration must be given to reducing the amount of materials used in all package designs. Whenever possible:

- Material usage should be minimized while optimizing package performance.

- Unnecessary materials should be eliminated.

- The need for secondary packaging should be questioned.

- Fully or partially recycled materials should be used, and the use of these materials should be identified on the package.

- Innovative packages should be designed that will help reduce volume for disposal or recycling.

Recyclability

Consideration must be given to determine if a package, or its components, can be recycled after use. This includes:

- Identifying the package material and whether it can be recycled.

- Determining if the package can be recycled in an existing solid waste stream.

- Using single versus multi-layer materials to aid in recycling.

- Determining if a common material can be used for all package components to aid in recycling.

- Providing for separation of packaging materials to maximize recycling potential and value.

Refillability/Reusability

Consideration should be given to the refillability/reusability of packages. This includes:

- Making the package compatible with larger, more material efficient, or smaller, concentrated product refill packages.

- Making elements of the package reusable with refill packages, such as pumps, sprayers, etc.

Disposability

Consideration must be given to determining the safe disposition of product packages. This includes:

- Determining whether materials can be landfilled or incinerated safely without forming hazardous decomposition products.

- Ensuring that the following heavy metals are not included in packaging materials: lead, chromium, mercury and cadmium.

- Determining whether materials are degradable and, if so, whether the by-products are safe for the environment.